Buried Alive

Other works by Jan Bondeson

The Prolific Countess

A Cabinet of Medical Curiosities

The Feejee Mermaid and Other Essays in Natural and Unnatural History

The Two-Headed Boy, and Other Medical Marvels

The London Monster

BURIED ALIVE

The Terrifying History of Our Most Primal Fear

JAN BONDESON

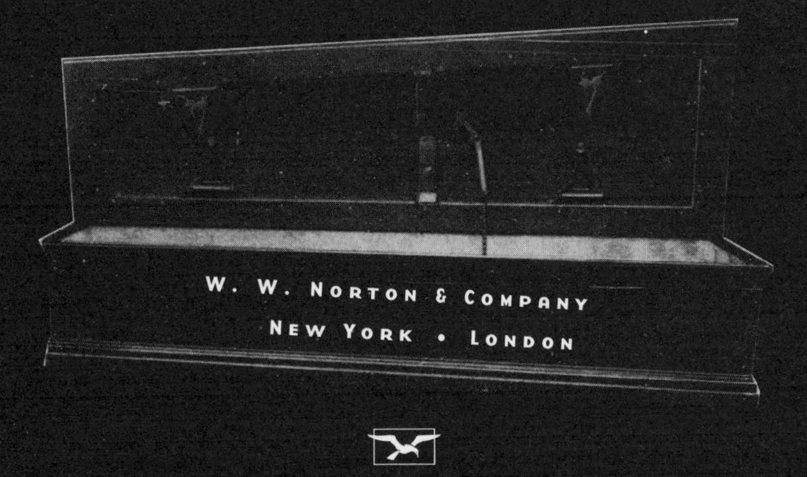

W. W. NORTON & COMPANY
NEW YORK • LONDON

Printed in the United States of America
First Edition

Financial support for this book from the Wellcome Trust (grant no. 063072) is gratefully acknowledged.

The text of this book is composed in Trump Medieval with the display set in Base Nine
Manufacturing by Maple-Vail Book Manufacturing Group
Book design and composition by Chris Welch

Library of Congress Cataloging-in-Publication Data

Bondeson, Jan.
Buried alive : the terrifying history of our most primal fear / Jan Bondeson.
p. cm.
Includes bibliographical references and index.
ISBN 0-393-04906-X
1. Burial, Premature. 2. Death—Proof and certification. I. Title.

RA1063 .B66 2001
616.07'8—dc21 00-048030

W. W. Norton & Company, Inc., 500 Fifth Avenue, New York, N.Y. 10110
www.wwnorton.com

W. W. Norton & Company Ltd., 10 Coptic Street, London WC1A 1PU

1 2 3 4 5 6 7 8 9 0

Contents

Introduction

> The unendurable oppression of the lungs—the stifling fumes of the damp earth—the clinging of the death garments—the rigid embrace of the narrow house—the blackness of the absolute Night—the silence like a sea that overwhelms—the unseen but palpable presence of Conqueror Worm—these things, with the thought of the air and grass above, with memory of dear friends who would fly to save us if but informed of our fate, and with consciousness that of this fate they can *never* be informed—that our hopeless portion is that of the really dead—these considerations, I say, carry into the heart, which still palpitates, a degree of appalling and intolerable horror from which the most daring imagination must recoil. We know nothing so agonizing upon Earth—we can dream of nothing half so hideous in the realms of the nethermost Hell. *—Edgar Allan Poe, "The Premature Burial"*

"There are certain themes of which the interest is all-absorbing, but which are too entirely horrible for the purposes of legitimate fiction." Thus Edgar Allan Poe begins his famous horror story "The Premature Burial," but his concerns about offending the delicate sensibility of his readers are by no means reflected in his own work. The sufferings of a person buried or immured alive form one of his favorite literary motifs, recurring in many of his stories.[1] What reader

An illustration for Edgar Allan Poe's "The Premature Burial" by the German artist Alfred Kabin. *From the author's collection.*

of Poe can forget the sufferings of the inappositely named Fortunato, who is walled in by his callous and implacable enemy Montresor, the horror of the teeth extracted from the senseless, sensuous corpse in Berenice, and the wild shriek from Roderick Usher that greets the appearance of his recently buried sister, dressed in her bloodied shroud and with evidence of a horrible struggle on her emaciated frame: *"We have put her living in the tomb!"*

Readers of these tales may comfort themselves with the notion that Poe must have exaggerated: surely people of the mid-1800s could not have been at risk of being buried alive? It has been speculated that Poe's obsession with premature burial was due to necrophilia or that he used it as a metaphor for loneliness—a sentient corpse buried alive by his unfeeling peers in the unfathomable tomb of the universe. Psychologists have favored the theory that he was actually longing for the maternal womb. There may be something in these hypotheses, but just like a writer of the 1920s featuring a sanatorium for consumptives, or one of the 1990s using AIDS as a metaphor, Poe related his fictional world to the medical fears and obsessions of his time. He was both inspired and horrified by the discussion about the risks of premature burial that raged in the mid-nineteenth century. Dozens, nay hundreds, of real-life tales of people who had either been buried alive or saved

An illustration for Poe's "The Fall of the House of Usher" by the German artist Alfred Kabin. *From the author's collection.*

from this dire fate at the last minute were on record in Poe's time. Stories about the horrors of premature burial could be found not only in medical journals but also in the popular newspapers and magazines, in Europe as well as in the United States. Compared with the flights of fancy enjoyed by some of the serious medical writers on the perils of premature burial, Poe was a mere dismal realist; compared with the *grand guignol* visions of horror they conjured up, his images were those of a squeamish maiden aunt.

In his *Thesaurus of Horror*, published in 1817, the English writer

John Snart tells the sad tale of a certain Mr. Cornish, who had twice been mayor of Bath.[2] This gentleman died, it was believed, and was duly buried, but after the earth had been shoveled down onto the coffin, a hollow groan was heard from underground. The coffin was speedily dug up and opened, and it was seen that the living corpse's knees and elbows had been beaten raw. The wretched man's life could not be saved, however: in his customary highly charged prose, Snart wrote that "he had drunk the bitter cup of *superlative misery* to the dregs!" Another of John Snart's cases concerns some gravediggers in Bermondsey churchyard in Surrey, who were exhuming a coffin for some reason or other. They were aghast to behold "a *master-piece of horror*!!!—A torn and bloody shroud! battered forehead! broken knees and elbows! (O God!! O God!! *misericordia*!!) all! all! appeared in view!! Even the barbarous prison of the *inanimate* coffin had, as it were, *relented*, and, relaxing its cruel grasp, suffering its screws to be torn out, to expose the *living inhumation* and the ensanguined consequences of the *horrid scuffle* . . . ."

The German medical practitioner Michael Benedict Lessing told another gruesome tale twenty years later.[3] A young Swedish girl, the only daughter of a wealthy ironmaster, had died in a state of advanced pregnancy and was buried in the local churchyard. In the evening, the verger heard pitiful groans from the grave he had filled with earth a few hours earlier. Coming closer, he thought he could discern the words "God! Jesus! Mercy!" He was a highly superstitious man and fearful of ghosts; not daring to examine the grave any further, he ran home to his cottage and pulled the bedsheets over his head. The girl was left alone in her dark coffin underground. It was not until the next day that the verger told the rector about the ghost he had heard. The clergyman was of course aghast at this belated discovery, reproached the superstitious menial for his tardiness in reporting it, and ordered the coffin to be speedily exhumed. As the coffin was opened, every person present gave a shriek of horror on seeing that the girl had given birth to a child in her coffin! Dr. Lessing cannot be accused of squeamishness, for he described every detail of the gruesome scene: the confused awakening in the narrow house underground, the pain and horror of the unnatural childbirth, and the girl's final torture in the unyielding coffin, writhing in blood and excrements.

Another, not less horrific instance comes from Whitehaven, Pennsylvania, and is dated July 1893.[4] The wife of a farmer named Charles Boger had died and was buried after death had been duly certified. Someone told Boger, however, that his wife had been subject to periodic hysteria before they were married and that there was a very real possibility of her having been buried alive. This terrible thought haunted the poor man, until at last he became a raving maniac. His friends then decided to reopen the wife's tomb to reassure him that all was well, but the spectacle that awaited them was not a pleasant one. The body was turned face downward, the robes were torn to shreds, and the glass in the coffin lid broken. The flesh was torn and the fingers were entirely missing; it was presumed that the wretched woman had gnawed them in her agony. It is not mentioned how poor Mr. Boger's mental equilibrium withstood this dreadful discovery.

These ghastly tales are just a sample of many hundred others in the same vein: in 1749, the French physician Jean-Jacques Bruhier reviewed 56 cases of premature burial or dissection, and 125 narrow escapes.[5] In 1905, the anti-premature-burial propagandists William Tebb and E. P. Vollum listed an impressive total of 161 individuals buried, dissected, or embalmed while still living, and 222 narrow escapes from the same dire fates.[6] Another alarmist, Dr. Franz Hartmann, to whom we are indebted for the pleasant story of Mr. Boger quoted above, claimed to have collected more than 700 cases.[7] There was speculation, from the number of skeletons found in lurid, unnatural positions upon exhumations of graveyards, that at least one-tenth of all human beings were buried before they were dead. The aforementioned Dr. Lessing wrote that if the tombs had been able to speak, a ghostly roar of accusation would have arisen from the coffins below ground, directed toward the careless and ignorant relatives who had suffered a multitude of still-living people to be prematurely buried: "You, you are to be blamed for this endless misery!"[8]

IN 1988, WHEN I was a young doctor, I first had occasion to read Philippe Ariès's famous book *L'Homme devant la mort*.[9] The vast learning and monumental scope of this groundbreaking study did not fail to impress me. I was particularly intrigued by the rather brief section about the eighteenth-century debate on the signs of death and the risk of premature burial, particularly as I had, already at that time, come across some rare nineteenth-century pamphlets on this subject, which complemented Ariès's account from another perspective. I decided to look into the medical and sociological literature on this subject, but information was not easy to come by. The huge modern textbooks on forensic medicine have nothing to say about the shameful past of medical science with regard to the certainty of the signs of death; they choose to ignore the fact that less than 150 years ago many medical practitioners freely admitted to being uncertain whether their patients were dead or alive. The fashionable historians of medicine have had little more to add. Not even the most massive textbooks, like Professor Roy Porter's *The Greatest Benefit to Mankind*, contain a single word about the debate over the uncertainty of the signs of death that raged throughout the eighteenth and nineteenth centuries. Even a specialist book like *Death and the Enlightenment*, by Dr. J. McManners, briefly dismisses the eighteenth-century fear of premature burial as a bizarre manifestation of French hypochondrical zeal.[10] But the fear of being buried alive was by no means confined to hypochondriacs; it affected not only France but the larger parts of the civilized world as well; it did begin in the eighteenth century, but lasted through the nineteenth century and well into the twentieth. Surely, it is of not negligible importance for the historian to know whether reliable techniques existed to determine whether people in extremis were dead or alive, not only because deplorable "accidents" like those related above are painful to contemplate but also because the vision of thousands of successful, complacent nineteenth-century physicians *burying their mistakes alive* tends to remove some of the luster from the medical breakthroughs of this era.

After some years of research, I published a brief essay about apparent death and premature burial in my popular book *Cabinet of Medical Curiosities*.[11] In 1998, I took part in an American TV documentary entitled "Buried Alive,"[12] and this experience prompted me to look through my old notes and to plan a full-length book on this subject. Much of my spare time in 1999 and early 2000 was spent on research in the British Library and the Library of the Royal Society of Medicine in London; I also traveled to Holland, Germany, France, Sweden, and Denmark in search of material and later made a prolonged visit to the United States.[13] This book is the first full-scale English-language work to address the history of the signs of death and the risk of premature burial.[14] It takes into account aspects of medicine, history, folklore, and literature; it deals not just with one country or period but attempts a wide-ranging, interdisciplinary view of its subject. Were the fears of Edgar Allan Poe's contemporaries justified, and did a mass burial of still-living people occur in the Victorian graveyards, as was asserted at the time, by many responsible medical writers, who could back their claims with impressive lists of gruesome case reports? Another unanswered riddle is what factors—medical, theological, historical, and sociological—caused the sudden upsurge in fear of apparent death and premature burial in the 1740s. How did the medical establishment react to the public fear of being buried alive, and what action did it take to safeguard against premature burial or to find more reliable signs of death? What people were active in the anti-premature-burial campaign, which lasted from the mid-eighteenth century until the early 1900s, and what were their motives?

The deep-rooted, primitive fear of being buried alive lies dormant in the complacent citizens of the late twentieth-century welfare state, but it does not take much to bring it out into the open: in Germany, there was a scare in the early 1980s, after some lamentable "mistakes" in diagnosing death had been played up in the newspapers.[15] The alarmist books inspired by this scare were very similar, in content and argumentation, to those that appeared in the 1740s, and the German medical establishment could easily brush them aside. But the question posed by these provocative books—whether the current criteria for declaring people dead are reliable, or if people are still at risk of being buried alive—is more difficult to ignore. Indeed, those who were present at the Selly Oak hospital in Birmingham, England, when sixty-five-year-old Michael McEldowney was declared brain dead in February 1974, would have felt deeply uneasy when answering this question.[16] When Mr. McEldowney's two kidneys were harvested on the operation table, his foot twitched, and then he started to breathe. . . .

Unlike Edgar Allan Poe, I make no promises that this book will not affect the nervous equilibrium of sensitive people: no subject is eschewed and no avenue of thought left unexplored. Nor will I exercise the caution shown by Shakespeare's *Hamlet*, as reinterpreted by a writer in the *Transactions of the Royal Humane Society* (the *prison-house* in this instance being the coffin):[17]

Oh Reader!—But that I am forbid
To tell the secrets of the *prison-house*
I could a tale unfold, whose lightest word
Would harrow up thy soul, freeze thy young blood,
Make thy two eyes, like stars, shoot from their spheres. . . .

I

Miracles of the Dead

In our graveyards with winter winds blowing
There's a good deal of to-ing and fro-ing
But can it be said
That the buried are dead
With their nails and their hair still growing?
—*Anonymous nineteenth-century limerick*

In classical antiquity, the absence of a heartbeat was the accepted sign of death.[1] The heart was the seat of life: the first organ to live and the last one to die. Breathing was considered just a regulator of the heat of the heart. There was an awareness that the brain influenced reason and sensation, but the brain's actions were still deemed dependent on the existence of a functional heart. Aristotle taught that a person was an integral combination of body and soul; the soul could not exist without a body, and the death of the body meant the death of the soul. He recognized three parts of the soul with different actions: the vegetative soul regulated bodily vitality, the animative soul controlled motion and sensation, and the rational soul, or the mind, governed the higher mental faculties. The rational soul might die without affecting the vitality of the body; indeed, animals could subsist their entire lives without one. The death of the vegetative soul always caused bodily death, however. Relatively little is known about what criteria of death were actually used in classical antiquity: one would presume that feeling the pulse had a central part, given the emphasis on the action of the heart as the divider between life and death. Immobility, coldness, and incipient putrefaction probably also played a role. Actually, when the classical physician spoke of "signs of death," he meant the physical signs in Hippocrates' *Prognostikon* that death was inevitable; the presence of these signs indicated that the doctor's work was done. According to the Hippocratic medical ethics, a doctor should then forecast the impending demise, collect his fee, and withdraw from the case. The actual diagnosis of death was left to the nonmedical attendants, often the patient's own family and relations.

There is evidence that already in classical antiquity some observers were aware that the criteria of death might sometimes be fallible. The seventh book of Pliny's *Natural History* contains a section on the signs of death among the Romans.[2] Rather pessimistically, Pliny wrote that the signs of death are innumerable, but that there are no signs that health is secure. Although many in number, these death signs were not always reliable. Shockingly, the consul Acilius Aviola and the praetor Lucius Lamia had both awakened on their flaming funeral pyres after being falsely declared dead, and the attendants could save neither of them from a most horrible death. Another Roman worthy, Gaius Aelius Tubero, managed to show signs of life while actually on the pyre, fortunately before it was too late. Pliny also mentioned several instances where people had been carried out on a bier to be buried, but returned on foot. He concluded, "Such is the condition of humanity, and so uncertain is men's judgment, that they cannot determine even death itself." Plutarch told of a man who had fallen from a precipice and who lay motionless for three days before returning to life as his friends carried him to the grave. In Plato's *Republic* can be found the tale of an Armenian soldier named Er who was slain in battle. Ten days later, the surviving soldiers returned to bury the dead and were surprised to find that even though all other bodies were corrupted, that of Er was still intact. This finding did not weaken their conviction that he was dead, however, and they put him on a funeral pyre, where, to the great surprise of all those present, he returned to life and was saved. The very influential Greek physician Galen recommended great caution in certain diseases, like hysteria, asphyxia, coma, and catalepsy, since the signs of life could be suspended for weeks without affecting the chance for recovery. In his *De locis affectis,* he commented on a case report given by Heraclides of Pontus, concerning a woman who had collapsed from "uterine suffocation" and was without a perceivable pulse or respiration for thirty days, before reviving. Galen was also aware that some people who died from excessive joy or grief were known to recover; moreover, he considered it unwise to consign to the grave too hastily those who had died from intoxication with alcohol or soporific drafts. In his *De medicina,* the influential Roman physician Aurelius Cornelius Celsus agreed, stating that the art of medicine was conjectural, and the signs of death not always totally reliable.[3] As evidence, he repeated the tale of Asclepiades of Prusa, who discovered that the "corpse" carried along in a funeral cortege was not really dead. But he also wrote that a sign should not be rejected if it was deceptive in just 1 out of 1,000 instances, if it held good in the other 999 patients. This controversy will recur again and again in the debate about the uncertainty of the signs of death. Celsus knew of cases where medical attendants had deserted their patients, after making a gloomy prognosis, only to find that the patients had recovered without their help. It was rumored, he wrote, that some people had even shown signs of life when carried to their funerals. There exists a manuscript of declamations attributed to Fabius Quintilien, but which is probably by a later Roman author, which contains some interesting observations concerning the reasons for delayed funerals in ancient Rome.[4] The pseudo-Quintilien wrote, "For what purpose do you imagine that long-delayed interments were invented? Or, on what account it is that the mournful pomp of funeral solemnities is always accompanied by sorrowful groans and piercing cries? Why, for no other reason, but because we have seen people return to life after they were about to be laid in the grave as dead."

Some interesting information about how the ancients viewed the reliability (or lack of it) of their procedures for declaring people dead can be found in some ancient Greek and Arabic novels and stories.[5] In the anonymous Greek novel *Apollonius, Prince of Tyre,* a physician finds a floating coffin with the corpse of a young woman in it. Some money has been put in the coffin, with a note stating that half of it should be used for a decent funeral pyre for the corpse, and the other half as a fee for the individual who found the coffin. The physician orders a funeral pyre to be prepared, but one of his apprentices actually manages to revive the presumed corpse by rubbing her body with ointment and oil; she makes her resurrection known by saying, "Doctor, please do not touch me in any way that is not proper. For I am the wife of a king and the daughter of a king." She turns out to be the wife of Prince Apollonius of Tyre, who was mistakenly buried at sea; it was equally fortunate and unrealistic that the coffin was not submerged by the waves. In a medieval Arabic tale, a baker eats a large meal of apricots and hot bread and falls lifeless to the ground shortly thereafter. The local doctor declares him dead from overeating, but fortunately the famous doctor Yabrudi passes by as the funeral procession is on the way to the burial yard. Yabrudi examines the baker and demands to know exactly what caused his death. He then prepares a powerful laxative, which soon has the desired effect; after a volcanic emptying of the bowels, the gourmandizing baker revives and is able to walk back to his shop.

A very long-lived and powerful literary motif, which will be encountered many times in this book, is that of the heroine who is buried alive, but saved by a robber.[6] The first incarnation of this legendary figure is to be found in the novel *Chariéas and Callirhoé*, by Chariton of Aphrodisias, a Greek writer active between the first and second century A.D. The heroine Callirhoé one day annoys her jealous husband, Chariéas, who is ungallant enough to kick her so violently in the stomach that she falls unconscious. She presents all signs of death and is promptly buried in a vault. Poor Callirhoé awakens in the tomb, however, breaks free from her shroud, and cries, "I am alive! Help me!" But no one can hear her, and she laments, "Alas, what misfortune! I am buried alive, through no fault of my own, and I will die a very long death!" But some pirates have decided to break into the vault to plunder it; they save Callirhoé and later sell her as a slave. The same theme recurs in the novel *The Ephiesians*, by Xenophon of Ephesos, a Greek writer active at about the same period. The heroine Antheia is to marry a certain Perilaos, but she considers this a fate worse than death and procures poison from a physician. On the morning of her supposed wedding day, she swallows it. The physician abided by the Hippocratic oath enough to be unwilling to participate actively in the suicide of one of his patients, however, and substituted a sleeping potion for the poison. Less valorously, the elderly practitioner did not mention this substitution to anyone, and when Antheia awakens, it is in the tomb. At that moment, however, a band of robbers breaks into the vault to steal her valuable jewelry, with which she has been buried, and she is saved from her premature burial and later reunited with her husband. These stories certainly imply that the fear of being buried alive, after having been mistaken for dead in a comatose state, and of awakening in the tomb, is a deep-seated one; the tale of the prematurely buried Callirhoé, and her pathetic lamentation in the dark and lonely vault, must have been listened to with a frisson of horror.

Much of the medical knowledge of antiquity was forgotten in medieval times, and there are fewer sources pertaining to the signs of death, and to the concerns about premature burial, from this period.[7] A remarkable anecdote tells that King Louis IX (Saint Louis) of France fell ill in 1244 with some kind of enteritis: he was severely weakened by the incessant diarrhea, and his doctors considered him to be dead. But when mass was said over the king's dead body, he moved and gave some other signs of life. Louis IX recovered completely, and in order to give thanks for what he regarded as a divine intervention to save his life, he equipped and led a crusade to Egypt.[8] Another curious story of uncertain origin (and doubtful veracity) tells that Thomas à Kempis, who died in 1471, was denied canonization because splinters of wood from the coffin lid were found embedded underneath his fingernails when the coffin was opened; why, if he had been worthy of becoming a saint, had he made such desperate efforts to postpone his meeting with his Maker?[9] The fact that some fourteenth-century English aristocrats, like Elizabeth de Burgh and John, duke of Lancaster, made bequests stipulating that their bodies be left above ground for several weeks without being embalmed, has by some observers been attributed to a fear of premature burial, but this was not mentioned in any of the actual bequests as the reason for these extraordinary delays.[10] It has also been suggested that uncertainty about the moment of death, and fear of a live burial, caused some medieval funerals to be long, drawn-out affairs, and that the religious custom of holding a candle near the mouth of the person in extremis was a crude test of death, since a breath would cause the flame to flicker.[11] Quite a few excavations of early cemeteries show that the most common burial positions were either with the arms by the sides of the body, the hands over the pelvis, or the arms crossed over the chest.[12] But there are both Danish and British examples of skeletons found in gruesome, contorted positions,[13] suggesting that the person was either killed violently and dumped into the grave face down or buried alive by accident or as a punishment.[14]

THE SEVENTEENTH CENTURY was the age of the old monster medicine, a time when medical science was immersed in a rich subculture of pagan myths, religious legends, and popular superstitions.

Death was defined as a state where life was extinct and the soul had left the body; a person could be either dead or alive, and no concept existed of a *process* of dying. Death was regarded as a wholly supernatural, obscure phenomenon, outside the limits of rational analysis. The resemblance between death and sleep was stressed: in both these states, the soul was considered to be concentrated outside the body, thereby capable of communicating with God. Another important observation concerned the mummies, which were believed still to have elements of life within them as long as the embalming preparations preserved them from corruption. Life was thus an exception from nature: a mystic force that could be retained in a cadaver by artificial means. The seventeenth-century scholars who wanted to study the phenomenon of death more closely did so by collecting observations concerning the dead body. They often preferred the popular, superstitious belief that the cadaver still had some degree of life and sensibility to the Christian dogma that the union and separation of body and soul accounted for creation and death.

At this time, curious anecdotes, legends, and folklore about the dead body abounded. Why did the body of a murdered man start to bleed profusely when the murderer entered the room, and why was a ghostly rattle of the bones from the tomb of Pope Sylvester II always heard just before the death of a pope? Why did the hair and nails of some cadavers keep growing after death, and why did some of them even cut new teeth? Who are the more happy, the dead or the living, and are all humans, even the most hideous monsters, resurrected in heaven? These were some of the questions addressed in Dr. Heinrich Kornmann's curious book *De miraculis mortuorum*, a treatise on the miracles of the dead, first published in 1610.[15] Among the sections on incorruptible saints, screaming corpses, speaking skulls, and jumping specters in Kornmann's book is to be found a brief note concerning a certain Cardinal Andreas, who died in Rome and was to be buried in a cathedral, where the pope and a body of clergy attended a service to honor his memory. But during the service the cardinal groaned and sat up in his coffin. This was looked on as a miracle and ascribed to the influence of Saint Jerome, to whom the cardinal was greatly attached. Another note describes the death of Archbishop Geron of Cologne, who was prematurely buried in a tomb in his own cathedral, and expired in the most lamentable manner; his sad fate was deemed just as miraculous as the phenomena discussed earlier.

Another view of the miracles of the dead is given in the German medical practitioner Christian Friedrich Garmann's similarly titled treatise *De miraculis mortuorum*.[16] Like many other seventeenth-century doctors and scientists, he was immensely erudite and preferred the compilation of numerous quotations from the classical literature, and the ancient repositories of curious medical anecdotes, to actual observations by himself or other contemporaries. The first edition of Garmann's book appeared in 1670, but he kept compiling more and more miracles, so the ultimate version of his book, posthumously published in 1709, was an encyclopedic work of more than twelve hundred pages. It provides a veritable dictionary of sixteenth- and seventeenth-century observations, beliefs, and superstitions about the dark underground world of cadavers.

The theme of the still-living dead is even stronger in Garmann's book.[17] Garmann gives examples of corpses that grew in size, moved, laughed, or wept, corpses whose facial expression changed, and even corpses whose hearts kept on beating. He cites instances of living children delivered from a dead mother by means of cesarean section, so it could be concluded that corpses could give birth. He also reports many observations of corpses with an erect penis. Indeed, when the corpses of dead soldiers were once undressed after a fierce battle, Garmann writes, many of them were in a state as if the engagement had taken place in the bedroom. The concluding forty-five pages of his book deal with the miracles of resuscitated and resurrected corpses. Garmann had read about a certain Zoroaster, who revived on the funeral pyre twelve days after being considered dead, and a boy described by a certain Dr. Valvasor, who came to life when his coffin was put into the grave. The learned Velschius told a story that was well known at the time: when grave robbers dug up the coffin of a recently buried woman in Cologne, she revived as they cut her finger to steal a valuable ring; exactly the same story was current about a young lady from Bohemia. Nowhere in Garmann's treatise is it concluded that these latter instances should perhaps prompt more careful scrutiny of individuals presumed dead to prevent the interment of still-living people; the events are interpreted as miraculous resurrections rather than as actual rescues from the tomb.

Another eerie phenomenon was the *masticatio mortuorum*, the ability of cadavers to eat their shrouds, or even their own fingers and arms. Garmann devoted an entire chapter to these matters: typically, a groaning sound is heard from the tomb, along with a loud smacking and chewing as the cadaver eats away. A dead woman had eaten her hands, and a man had devoured his entire body. There was no question that these noisy cadavers were actually still alive, and Garmann quoted observations where people had actually dug up the coffin in question and found that its groaning and chewing inmate had been reduced to a loathsome heap of putrefaction. The *masticatio mortuorum* was considered to be a bad omen, a warning that famine or disease was imminent. In 1734, the German doctor Michael Ranft published an entire "Treatise concerning the Screaming and Chewing of Corpses in their Graves." He agreed with Garmann that this phenomenon was a supernatural one, which could not be explained by medical science.[18]

The concept and application of the signs of death did not change much from the time of Celsus until the mid-seventeenth century: the heartbeat and arterial pulsations were the central criteria, but various crude tests for respiration and sensibility also existed. Many people were not seen by a doctor during their final illness, nor were they examined after death; the practical application of the signs of death was thus often left to laypeople. Some sixteenth- and seventeenth-century physicians were aware of the dangers of hasty funerals. Already the old Salmuth recommended caution in these matters; in particular, the bodies of women known to be of a nervous and hysterical disposition should be left above ground for three days before burial. Petrus Forestus, another sixteenth-century medical writer, agreed, since he actually knew of cases where women "buried during a paroxysm of the hysteric passion" had returned to life in their graves.[19] In 1670, the Germans Theodorus Kirchmaier and Christophorus Nottnagel pointed out the difficulty in distinguishing real from apparent death, and the wisdom of delaying the funeral for some days when there was any doubt.[20] Cessation of heartbeat and respiration, coldness and insensibility were but uncertain signs of death and should be judged with caution. They had spoken to a man from Wittenberg named August Schwenske, who had been mistaken for dead when just three years old, but had been rescued and was now a grown man and a father. They quoted the opinions of some learned contemporaries with approval: Amatus Lusitanus had observed a girl from Ferrara who had been presumed dead from apoplexy, but her mother was so distraught that she let no one bury her. On the third day after the apoplectic attack (probably a lengthy epileptic seizure), she revived. The most famous instance of premature burial involved the soldier François de Civille, who was said to have been thrice declared dead and as many times rescued from the tomb. He had been born by caesarean section to a dead mother exhumed from her coffin. He became an army captain. Later he was severely wounded in the siege of Rouen in 1563 and buried alive in a common grave on the battlefield. His servant, who had wanted to dig his master a more fitting grave, discovered that he was not dead. While François de Civille was recovering from his wound, a troop of hostile soldiers burst into the house and threw the convalescent into a dung heap, where he remained buried for three days until dug out a third time, and nursed back to health. According to a gravestone in Milan, François de Civille was finally buried in the graveyard of that city; he had died at the age of 105, from a chill contracted while "serenading the lady of his heart all night long."

Giovanni Maria Lancisi, first physician to Pope Clement IX, had once seen the presumed corpse sit up in his coffin during the funeral mass in a church in Rome, and this naturally made him doubt the current principles for declaring people dead. In a treatise on sudden death, published in 1707, he pointed out that errors such as this one were due not to deficiency of the medical art but to the carelessness of certain practitioners, who abandoned their patients to the mortuary bearers too early.[21] He suggested an impressive list of tests to be used in situations when there was uncertainty whether life was extinct: smelling salts, sneezing powder, tufts of wool to be put into the nostrils and a dry mirror put before the mouth to test for respiration, and a full goblet of water on the chest to test for motion of the diaphragm. The pulse was of paramount importance as a sign of death, but the physician should take care not to feel his own pulse by mistake, and thus declare a corpse to be still living.

THE FIRST EVIDENCE of any concerns in England over the signs of death and the risk of premature burial appears in Sir Francis Bacon's *Historia vitae et mortis*.[22] He knew that the famous philosopher John Duns Scotus had been liable to some obscure kind of fits, during which he became surprisingly deathlike. When he was staying in Cologne, he was declared dead and buried after a fit, but his servant came up and said that the wretched man had probably been buried alive. He, the servant, had been instructed to prevent this at all costs, but he had taken a detour and arrived too late. When Duns Scotus's coffin had been exhumed, it was seen that the corpse's hands were torn and the fingers gnawed, from which it was concluded that the servant's apprehensions had been only too valid. For many years, Duns Scotus's tomb had a plaque with a Latin inscription, which was translated as follows:

Mark this man's demise, o traveler,
For here lies John Scot, once interr'd
But twice dead; we are now wiser
And still alive, who then so err'd.

The seventeenth-century medical and scientific literature provides little additional evidence that the English people of this time worried about premature burial. The subject of people buried alive occurs in some popular pamphlets, but these publications were notoriously fanciful and should not be taken as proof that the fear of apparent death and premature burial was widespread at this time. One of them, *The Most Lamentable and Deplorable Accident*, tells the sad tale of Lawrence Cawthorn, a butcher in Newgate Market in London, who suddenly fell ill sometime in 1661.[23] His wicked landlady, eager to inherit his belongings, saw to it that he was hastily buried. But at the chapel where Cawthorn was buried, the visiting mourners were horrified by a muffled shriek from the tomb and by a frenzied clawing at the coffin walls. When finally disinterred, Cawthorn's lifeless body was a horrid sight: the shroud was torn to pieces, the eyes hideously swollen, and "the brains beaten out of the head." It was concluded, "Amongst all the torments that Mankind is capable of, the most dreadful of them, and that which Nature most shrinks at is to be buried alive," and the covetous landlady was roundly accused of having deliberately put the butcher living into the tomb. According to another pamphlet, entitled *A Full and True Relation of a Maid Living in Newgate Street*, attempts at rescue were similarly futile when a sixteen-year-old girl was heard to groan and cry from her four-day-old grave in a London cemetery.[24] No seventeenth-century pamphlet was complete without a moralistic conclusion: the poor girl's master and mistress had abused her so horribly that she had been overheard to pray that she would rather be buried alive than live in such misery; the Almighty did not tarry long before fulfilling this imprudent wish.

Much more sinister, and also more truthful, than either of these two pamphlets was *News from Basing-Stoak*, published in 1674, which heralded what was considered one of the most celebrated cases of premature burial of all time.[25] Madam Blunden, a native of Basingstoke unflatteringly described as "a fat gross woman who liked to drink brandy," one evening felt indisposed and ordered some poppy water from the apothecary. She drank most of it and fell into a deathlike stupor. Her servants sent for the apothecary who had prepared this decoction of opium, and after surveying what was left in the bottle, the apothecary pronounced that she had taken enough not to wake up for forty-eight hours and would therefore never rise again. Madam Blunden's relations and servants were convinced by this dubious deduction on the part of the obtuse medical attendant. Her husband, the wealthy maltster William Blunden, one of the leading citizens in Basingstoke, wanted to defer the funeral until he could return from London, but the vile smell from Madam Blunden's huge body was so overpowering that her relations unanimously decided to have her buried the day after her presumed demise. As the coffin was set down between two stools, one of the pallbearers was heard to joke that they had probably made Madam Blunden's coffin too short, since he had clearly seen her stir because she could not lie easy. The man was rebuked for his levity.

Two days after the funeral, some schoolboys were playing in the burial ground near the Chapel of the Holy Ghost, when they heard a hollow voice emanating from the earth near Madam Blunden's grave. Coming nearer, they could hear the plea "Take me out of my grave!" intermixed with "fearful groans and dismal shriekings." Terrified, the boys ran to fetch their schoolmaster, but the brutal pedagogue took no action except to reproach them severely and to thrash some of them for telling such obvious lies. The morning after, the boys were back in the churchyard and heard the same ghostly voice from underground. This time, the usher did not resort to his birch rod, but uneasily suspected that there might be something in this extraordinary story after all. He went to ask the verger to have Madam Blunden's grave opened, but this individual refused to do anything of the kind without permission from the churchwardens. This body met the same afternoon and discussed the matter at length; not until the evening was Madam Blunden finally exhumed. The body being lamentably bruised and beaten, it was presumed that the injuries were self-inflicted during the horrid struggle underground. No signs of life could be detected, but the churchwardens nevertheless posted some custodians to stand watch over the grave during the night. It was a wet night, however, and these custodians left the body in the coffin, put the lid on, and went indoors. The next morning, it was seen that Madam Blunden had again revived: the winding sheet was torn off, and she had scratched herself in several places and beaten her mouth until it was covered with blood. A doctor was called, but he could only confirm that all life was gone, this time for good. There was of course an inquest after these almost unparalleled atrocities, and several individuals were held responsible for Madam Blunden's death. But after a physician of the town (perhaps the same unwise apothecary who decanted the poppy water in the first place) had testified under oath that he had applied a looking glass to her mouth without being able to discern any breath coming from her, they were let off. The town of Basingstoke was made to pay a large fine, however, for this neglect.

In 1819, the independent Minister Joseph Jefferson made inquiries in Basingstoke whether any person alive could remember the dreadful fate of Madam Blunden.[26] Two old ladies recalled that their ancestors had been among the schoolboys involved; they both used to say that they had heard a noise in the vault and that the bruises on Madam Blunden's face and the dew inside the coffin had led people to conclude that she had been buried alive. Mrs. Paris, the local midwife, was blamed for having persuaded Mr. Blunden to bury his wife too hastily; the servant maid Ann Runnegar, who had handed her mistress the fatal poppy water, had lost her reason on account of the dreadful event. The ancient Holy Ghost Chapel in Basingstoke was in ruins already in Jefferson's times, but these ruins are still standing. It is known that Madam Blunden was finally buried in the Liten burial ground close to the chapel. In 1896, the burial reform propagandist William Tebb visited the Blunden vault, which he was in some way able to identify although the inscription on the gravestone was completely obliterated.[27] When I made the same pilgrimage to these interesting old ruins in September 1999, I was less fortunate in this respect, since all remaining older gravestones have been defaced by the combined actions of time and the elements. It is interesting to note that there is still a local tradition in these parts that, a long time ago, a woman was buried alive in this cemetery and that the place is haunted.

IT IS EASY enough to quote the seventeenth-century medical works on the subject of the uncertainty of the signs of death, but more difficult to determine what impact the learned and obtuse discussion in these dusty tomes really had on the attitude of laypeople. There are some interesting French data concerning this, however, gleaned from studies of early wills.[28] After all, it was a natural thing for people fearing premature burial to insert a clause in their wills to make sure they escaped this dire fate. The earliest example of this is offered by Princess Elizabeth of Orléans, whose will dated March 1, 1684, explicitly requested that she not be shrouded until twenty-four hours after death and that the body then be cut twice under the soles of the feet with a razor. The will of the parson Jean Poitevin, dated 1705, was even more elaborate. The parson desired that, at the chosen hour for God to call him up to his Heavenly Kingdom, his dead body be well taken care of. It should be carefully determined that he was really dead, however, by means of the most certain tests of death (he made no suggestions himself). This was not, he stressed, because he was greatly attached to being alive, but because he had read about people being buried alive and knew that this had happened to one of his ancestors. The wills that specified any fears of premature burial were a small minority, however. An investigation of one thousand Parisian wills from 1710 to 1725 turned up only two that had clauses about avoiding premature interment: again, a townswoman requested to be cut under the soles of the feet before burial, and the widow of a marquis asked to be kept above earth for twenty-four hours before burial and to have her chest opened so that the heart could be seen, thus making a live burial impossible.[29]

In the seventeenth-century, many physicians were aware of the risk of fatal errors in declaring people dead during plague or cholera epidemics. The epidemics of the time usually caused mayhem: the carnage of victims was almost unbelievable, and the authorities often tried to limit the spread of the disease by burying the victims quickly. These conditions did not allow for any careful scrutiny of each dead (or perhaps not so dead) body that was hurled into the mass graves. The chronicles of Simon Goulart tell that when the town of Dijon was ravaged by the plague in 1558, people died so quickly that they had to be buried in plague pits instead of individual graves. The supposedly dead body of a woman named Nicole Tentillet was thrown into one of these pits. She revived the morning after and made efforts to climb out, but the weight of the corpses above her held her down. Four days later, the gravediggers brought more corpses and fortunately observed Mme Tentillet's predicament; she was taken back to her own house, where she recovered completely.[30] The influential papal physician Paolo Zacchia saw a young man who had twice been believed dead from the plague in Rome in 1656, but both times recovered. Zacchia presumed that several people had been erroneously buried alive during this epidemic. He recommended that initial decomposition and a fetid smell, heralding initial putrefaction, should be awaited before burial in doubtful cases.[31] In his treatise on the plague, the Dutch physician Isbrand van Diemerbroeck agreed. He had once seen a peasant named Pierre Petit, from the village of Bommel, not far from Nijmegen, who had been declared dead in a plague epidemic, but revived after several days of a deathlike stupor. Van Diemerbroeck wrote that it was very likely that people had been buried alive in plague epidemics, particularly because it was customary to bury people just a few hours after death.[32]

The great plague of Marseilles raged from 1720 until 1722 and claimed about fifty thousand victims, half the city's population.[33] The streets were cluttered with corpses, and mass graves were dug all around the town. One of them, at the Observance monastery, was excavated in 1994 when a housing development was being built. The excavation process brought a curious discovery: in two of the corpses discovered, an inch-long bronze pin was found in contact with the big

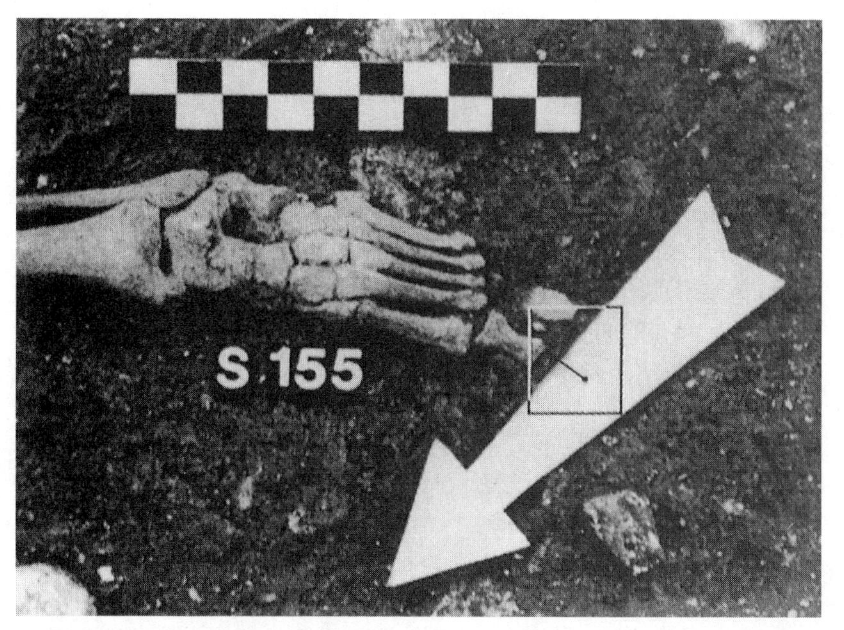

An illustration showing evidence of pin implantation in the big toe as a method of verifying death during the great plague of Marseilles in 1722. *From the article by Leonetti et al. in the* Journal of Forensic Sciences *42 (1997): 744–48, reproduced by permission.*

toe, in a position as if it had been deliberately driven underneath the nail of the big toe.[34] Pin implantation under the toenail as a means of verifying death had been suggested by both Zacchia and Lancisi in the books previously quoted, but this is the first historical evidence of the actual use of this method.

England saw similar concerns about premature burials during the 1604 and 1665 plague epidemics.[35] Persons believed to be profiting from the disease incurred much hatred and distrust, and doctors as well as keepers of the sick were accused of killing people or deliberately burying them alive. In the 1665 epidemic, a butcher in Newgate Market thought to be dead was not taken away from his room, because of the negligence of the corpse bearers. He returned to life during the night, came downstairs, and complained to a little girl that he felt cold.[36] Another story tells of a poor simpleton piper who had passed out in the street after having had too much to drink in a public house. When the corpse cart came by to pick up a corpse from the house next door, the men threw the piper's lifeless body onto the cart as well, believing him to be yet another plague corpse. One version of the tale has the piper make his resurrection known by playing his instrument! Daniel Defoe spoke to the man who drove the cart, however, and found out the truth. The unconscious piper, buried under a heap of corpses, had suddenly shouted, "Hey! Where am I?," when the cart was unloaded at the Mount Mill plague pit. The men were frightened, believing that he was a ghost, but one of them said, "Why, you are in the dead-cart, and we are going to bury you!" The simple piper replied, "But I an't dead though, am I?," which made them laugh.[37] The poet William Austin spoke for many when he wrote,[38]

Wisely they leave graves open for the dead
'Cause some too early are brought to bed.

One out of trance return'd, after much strife,
Among a heap of dead, *exclaims for* life.
One finding himself as some maid *hard lace't,*
Or as a watch *for pocket, straightly case't,*
Equaly terrifi'd with pain *and* fear,
Complains to those can neither speak nor hear.

II

The Lady with the Ring and the Lecherous Monk

The Doctor he sat by Frau Richermondt's side
With his anti-epidemical
But the fearful disease permitted no ease
And baffled each drug and chemical

So he rose from his chair with a solemn air
And thrice he shook his head;
And he said, with a sigh (that was all my eye),
"Frau Richermondt is dead!"

Thus begins the humorous poem "A Tale of Old Cologne," anonymously published in the *Notes and Queries* of 1882, after another correspondent to this periodical had brought up the subject of an ancient example of a miraculous rescue from premature burial, the tale of Frau Richmodis von Aducht.[1] In the Neumarkt, an ancient square in the old town of Cologne, lived the patrician family von Aducht: Menginus and his wife, Richmodis.[2] Their house was known as Die Papageien, since a sign with the family arms, three crowned parrots, was prominently displayed. In 1357, the legend tells, Frau Richmodis became indisposed and died after a short period of illness. She was buried in state, and her grieving husband had allowed a valuable gold ring to remain on her finger:

Snug under the Dom, in a stately tomb,
Frau Richermondt is laid;
The glare of the day has passed away,
And a mass for her soul has been said.

'Tis deep midnight, when the noisome sprite
Plays leap-frog over the grave.
What man, then, is here, who seems not to fear
The wrath of the dead to brave?

'Twas the sexton, and few were the words he spoke
And he breathed not a word of sorrow;
But he said "If I cabbage that ring to-night,
I shall be all the richer to-morrow!"

Armed with a spade, a crow, and a lantern, the sexton set out to dig up her coffin. He hauled the coffin out of the grave and forced the lid open. But when the sacrilegious robber tried to pull the ring off the corpse's cold finger, he was unable to move it over the knuckle and drew a knife to cut the finger off:

But scarce was the blade to the cold skin laid
Ere—horror of horrors to tell!—
The death-cold claw grasped the sexton's paw,
And the knife from his clutches fell.

Well, what did he do? Just as I, Sir, or you
Would have done in a similar plight;

In a swoon he fell down on the cold, clammy stone,
And in falling banged out his light.

Frau Richermondt rose—not so dead, you'll suppose,
As it seemed when she was buried—
Hit the sexton a blow on the shins with his crow,
And into the city she hurried.

After this fortunate escape from the tomb, Frau Richmodis picked up the sexton's lantern, which she used to light her way to the family house in the Neumarkt. She knocked on the door again and again, until finally a servant came to open it. But the superstitious menial, fearful of ghosts, fled yelling when he saw the shrouded figure of his former mistress standing outside. Herr von Aducht was called, but he, too, was very frightened:

Herr Richermondt groaned, and Herr Richermondt moaned,
And he uttered a hideous scream
For facing him there stood a ghastly nightmare
Intruding herself on his dream.

"Oh goblin immortal, avaunt from my portal,
Back, back to your region so drear!
For my wife I have lost her, thou ghostly impostor,
She's there—so she can't be here."

Frau Richmodis pleaded with her incredulous husband and the servants to be let indoors, but they were convinced she was a ghost. Herr von Aducht even said it was just as impossible that his wife left her tomb and returned to her house as it would be for his horses to leave their stables and run up the back stairs to the top floor of the house. But then a second miracle took place:

The words were scarce spoke when a wild neighing broke
On the silence of the night;
At the window, I ween, two horses were seen
In the glimmering, pale moonlight.

Herr Richermondt saw, and he crossed himself o'er
From the tip of his nose to his chest,
Then seizing his wife in her newly found life,
He clasped her once more to his breast.

THE TALE OF Richmodis von Aducht is certainly very old. The Koelhoff chronicle of Cologne, published in 1499, mentions that a wife in that city had been prematurely buried and disinterred alive by a gravedigger who had wanted to steal her ring. The year of this memo-

Reichmuth von Adolch rises from the tomb in this German print, engraved by Johann Bussenmacher in Cologne, 1604. In the background can be seen the funeral procession (*right*) and the resuscitated woman's attempt to gain entry into her house (*left*).

rable happening is 1400, and although the Papageien house in the Neumarkt is mentioned, the name Richmodis von Aducht is not. The next mention of the legend is in 1582, when Dr. Johann Ewich described a painting in the Apostelkirche in Cologne that depicted the miraculous return from the grave of an (unnamed) woman in that town. In 1604, the publisher Johann Bussenmacher issued a print of the original painting in the Apostelkirche, with some additional designs; he also gave a much more complete account of the memorable events of 1357. Richmodis von Aducht, the wife of a local magistrate, was declared dead after a plague epidemic. Rescued from the tomb and taken back by her startled husband, she stayed with him for seven more years and gave birth to three sons, all of whom took holy orders. She was known for her piety and industry, and wove an immense quantity of linen, some of which could still be seen, suspended next to the painting on her tomb in the Apostelkirche. Bussenmacher also appended a poem, of which I render a partial translation:

When with that ring they buried her, the husband's tears were running,
But the gravedigger took note of it with malice and with cunning,
And when the shade of night came down, as sure as it came he
With 'prentice lad and lantern, to help him dig and see.

They hoped the coffin case to haul out of the churchyard ground
And that the golden ring could still by them be found,
But when the lad the casket lid was raising with his knife,
The lady in the coffin sat up and came to life!

Thanks to several reissues and re-engravings of the Bussenmacher print, the memorable tale of Richmodis von Aducht became well known throughout Europe. It was mentioned in Simon Goulart's popular *Histoires admirables et mémorables* and in many other historical chronicles and repositories of memorable anecdotes. But it is worthwhile to examine two other, independent sources of the same legend. In 1688, the French traveler Maximilien Misson came to Cologne, where he visited the Apostelkirche and saw the painting and the linen exhibited there.[3] When he made inquiries, he was told that the miracle of Richmodis von Aducht had occurred in 1571 (!). The utilitarian German who told the story added the happy ending that Herr von Aducht and his wife had "made Machines necessary to let down the Horses." He also went to the Neumarkt, where he saw two sculpted wooden horses' heads emerging from the upper windows of an old house. Baron Risbeck, another traveler, was even more incredulous. He considered the entire story just another example of the superstition and credulity of the good burghers of Cologne.[4]

In 1785, the part of the Apostelkirche containing the painting of Richmodis von Aducht was demolished. The ancient linen that had been kept in the church for more than two hundred years was saved and put on display for the many credulous people who wanted to see this relic of the Lady with the Ring. But it was later found that it was a late twelfth-century fasting cloth and had nothing whatsoever to do with the legend of Richmodis von Aducht.[5] Another German scholar discovered that there had really been a Richmodis von Aducht (or von der Aducht); she came from the noble family of von Lyskirchen and was married to Waltelmus Mengin von Aducht just as the legend told. The problem was that her husband had died before 1342 and that the will of Richmodis von Aducht, which was still kept in the Cologne City Archives, specified that she was by this time (1346) in late middle age and had eight children alive. Moreover, the von Aducht family lived not in the Papageien house but in quite another dwelling in the old section of Cologne.[6] These observations make it unlikely that Richmodis von Aducht was really the Lady with the Ring who had almost been buried alive in the fourteenth or fifteenth century.[7] Some scholars have concluded that the story is a complete fabrication, but it should be noted that since there was a particularly severe plague epidemic in Cologne in 1349, one of the presumed corpses might have revived in the cemetery, thereby providing the catalyst for this remarkable legend, the name Richmodis von Aducht, the story of the robber, and the ludicrous episode of the horses all being later additions. As was noted in the preceding chapter, the theme of the prematurely buried heroine rescued by robbers who want to steal valuables from the tomb was known already in the second century A.D. At any rate, Richmodis von Aducht has obtained an undeserved fame before posterity, and the story of the Lady with the Ring is still told in Cologne. A street near the Neumarkt was named Richmodis-Strasse after her in 1877, and from the upper windows of what is nowadays called the Richmodis house, the two sculpted horses' heads still look out onto the busy square below.

WHEN BARON RISBECK declared himself to be a firm unbeliever in the tale of the Lady with the Ring in 1785, his main argument was that he knew that exactly the same story was current in several other German towns. In 1920, the German ethnologist Dr. Johannes Bolte made a detailed survey of the legend and was amazed to find that not fewer than nineteen other German cities—Hamburg, Dresden, and Lübeck among them—had a Lady with the Ring story of their own.[8] Often, the Lady was named and additional details given; in at least eleven instances, horses' heads were on display as memorials of the ludicrous ending of the tale. In Freiburg, a fasting cloth allegedly woven by the heroine was on prominent display, just as in Cologne.

The fame of the Lady with the Ring does not end there. A story well known among the English aristocracy in the nineteenth century was that of Emma, countess of Mount Edgcumbe.[9] Like Richmodis von Aducht, she was a real person: born in 1729 and married in 1761. With her husband, the earl, she lived at the family mansion of Cothele House in Cornwall. The legend tells that she was erroneously declared dead, and buried in the family vault, but rescued by a thieving sexton who broke into the vault to steal her valuable rings. Dressed only in her shroud, she slowly and painfully made her way back to Cothele House, along a path that was ever after known as the Countess's Path.

Another Lady with the Ring is awakening from her death trance beside her open coffin; the corpse of the inevitable sexton, who has been struck dead with fright, is lying beside her. One of the curious illustrations in *The Uncertainty of the Signs of Death*, the 1746 English edition of Bruhier's book. *Copyright British Library.*

The earl of Mount Edgcumbe, just as startled as the lugubrious Herr Richermondt, let out a shriek when he saw her, believing her to be the ghost of his wife. He refrained from uttering any silly oaths about horses running upstairs, however, and simply went down to let his wife indoors. When Dr. Robert Wilkins, author of *The Fireside Book of Death*, examined the story further, he could find no record of the death and funeral of the countess in the relevant parish archives; the path from the family vault to the house was ten miles long, and would thus have been quite a strenuous walk for a lady who had just awakened from a death trance. Furthermore, exactly the same story was current about one of the countess's forebears, Lady Anne Edgcumbe, who had lived in the preceding century.[10] Similar legends were current in several other noble families. What is purported to be the oldest of all is also from Cornwall: the fourteenth-century tale of Annot of Benallay, which was the subject of a poem by the Reverend Robert Stephen Hawker, containing these lines:[11]

They shrouded her in Maiden-white,
They buried her in pall,—
And the ring he gave his troth to plight,
Shines on her finger small.

"Shame! Shame! Those rings of stones and gold,"
The ghastly caitiff said,—
"Better that living hands should hold,
Than glitter on the dead."

The evil wish wrought evil deed—
The pall is rent away,—
And lo! Beneath the shattered lid
The Flower of Benallay.

But life gleams from these opening eyes!
Blood thrills that lifted hand!
And awful words are in her cries,
Which none may understand!

There are numerous other English Ladies with the Ring.[12] One was Lady Katherine Wyndham, the mother of the first earl of Egremont, who was entombed in the family vault at St. Decumans in Somerset; another was Mrs. Hannah Goodman, wife of the Reverend Richard Goodman, vicar of Ballymodan, Bandon. The memorial tablet to Constance Whitney in the Church of St. Giles without Cripplegate in Lon-

The interior of the Church of St. Giles without Cripplegate, in London, a drawing by John Carter in 1781. To the left can be seen the monument to Constance Whitney, one of the several British incarnations of the Lady with the Ring, who is arising from her tomb. *Copyright Guildhall Library, Corporation of London.*

don, representing her rising from her coffin, was said to commemorate her rescue from the tomb through the cupidity of a sexton, although rationalists objected that it was more likely to represent the pious hope of her resurrection in heaven, as supported by the inscription underneath. Miss Wynne's *Diaries of a Lady of Quality* name a certain Mrs. Killigrew as the Lady with the Ring, and similar legends were current in the families of Longstone of Longstone in Derbyshire, and the Barons St. John in Bedfordshire. In Scotland, there were Lady with the Ring stories about Marjorie Elphinstone, who lived in Ardtannies in the early seventeenth century, and about Margaret Halcrow Erskine, mother of the founders of the Secession Church of Berwickshire, who was widely reported to have "died" in 1674, only to be rescued by the usual thieving sexton out to steal her ring.[13] Northern Ireland has its own Lady with the Ring: Marjorie McCall from Lurgan, County Armagh, who lived sometime in the early eighteenth century. The story is well known locally, and a tablet on Mrs. McCall's gravestone still bears the inscription "Marjorie McCall—lived once, buried twice."[14] The fishing village of Lunenburg in Nova Scotia, Canada, has yet another Lady with the Ring legend, which probably originated with the German settlers who arrived there in the eighteenth century.[15]

The tale of the Lady with the Ring also occurs in Scandinavian popular mythology. In Sweden, five variants of the traditional tale exist: for example, several works of local history claim that the mother of the master mechanic Carl Jacob Hublein, of Karlstad, was saved from the grave by a thieving gravedigger.[16] Two other versions of the legend have a special twist. In the more benign form, the Lady with the Ring chokes on a piece of meat and is declared dead. The inevitable sexton digs her coffin up and, to pull her ring off, puts his knee on the chest of the presumed corpse. The piece of meat is then forced up the gullet, and the Lady revives; the magnanimous husband gives the sexton the ring as a present for saving his wife's life. In the darker version, the ruthless Scandinavian grave robbers show that they are made of sterner stuff than their wimpish French and German counterparts. Far from fainting or dying from shock when the Lady with the Ring revives in her coffin, they beat her to death with their spades, steal her jewelry, and bury her for good this time.[17]

IN THE SEVENTEENTH and early eighteenth centuries, the *historia valde memorabilis* was a peculiar phenomenon: oft-repeated, memorable stories with a moral that existed in many variations. One example is the legend of the noblewoman who insulted a poor beggar woman with twins: as the result of a curse, the noblewoman herself gave birth to 365 children, as many as the days in a year. The original version of this legend concerned Countess Margaret of Henneberg, who died in Loosduinen, Holland, in 1276, but there were several later versions of this legend, all with different protagonists from the European nobility.[18] Then there was the memorable story of the evil bishop in the mouse tower: in the year 913, Bishop Hatto of Mainz deliberately burned a barn full of starving people, and as a suitable punishment an army of rats marched toward the bishop's palace. Although Hatto repaired to a mouse tower built on a steep rock in the Rhine to serve him as a refuge in such an emergency, the rats swam the Rhine and devoured the prelate there. Again, this tale was repeated, with relish, about many other evil persecutors of the poor. The legend of the Lady with the Ring is another of these memorable stories; it spread from fifteenth-century Germany to become well known, with different names and dates attached, in almost every European country. Since it lacked the wholly ludicrous elements of other widespread stories about bishops eaten by mice and girls born with pig's heads as a divine punishment to the mother, the tale of the Lady with the Ring survived their gradual demise in the eighteenth century, and many variations of it were still current in late nineteenth-century folklore.

Nor was the Lady with the Ring legend the only popular tale about swooning ladies being saved from the tomb. There existed several legends on this theme, all in many variations, and the great majority can be presumed to be complete inventions. However, they are of considerable importance, since these idle tales were uncritically repeated by most later writers on premature burial, from the somber Latin pages of the doctoral thesis to those of the vulgar anti-premature-burial pamphlet. The oldest, and most famous, was the legend of the Lady and the Ring, closely followed by another variation on the same theme, known at least since the fourteenth century: the legend of the Two Young Lovers.[19] They are forbidden to marry by the young lady's stern father, who prefers a wicked nobleman (or tax collector) as a son-in-law. Filial piety makes her go along with this scheme, and she jilts her childhood sweetheart. But she cannot stand being separated from him and is unhappy in her marriage. She gradually falls into a decline and finally dies from the effects of a broken heart. Her grief-struck ex-fiancé visits her tomb in the family vault and plans to cut his throat with a razor there, but the girl awakens from her deathlike trance just in the nick of time to stop him. In one version, they visit the wicked nobleman, and he drops dead on seeing the presumed ghost of his wife. The French tax collector, the villain in the other version, is made of sterner stuff: not only does he survive the shock of seeing his wife return from the dead, but he actually tries to reclaim her as his lawful wife, thereby forcing the young couple to escape to England.

The two young lovers are reunited, after the lady has been rescued from her premature tomb. Another illustration in the 1746 English edition of Bruhier's book. *Copyright British Library.*

An early and oft-repeated version of the tale of the Two Young Lovers takes place in Florence during the great plague of 1400, which circumstance also provides a good excuse for the hasty burial of the heroine, Ginevra de Amieri, to the despair of her faithful lover, Antonio Rondinelli. Interestingly, there is some overlap between this early version and the Richmodis von Aducht story. Just a few days after receiving the wedding ring from the man she detests, but is forced to marry, Ginevra "dies" from the plague and is buried alive. She awakes in the tomb and forces her way out, returning to her lawful husband as a dutiful wife should, along a path that was long sentimentally revered in Florence and called the Way of Death. Just like Richmodis von Aducht's husband, poor Ginevra's spouse is a superstitious soul and shuts his door against the shrouded "ghost" that demands entry. At the houses of her father and uncle, she is also treated with horror and dismay. She then seeks out the faithful Antonio Rondinelli, with the agonizing cry "I am no spirit, Antonio! I am that Ginevra you once loved, but who was buried—buried alive!" They are blissfully united and later marry in the cathedral of Florence, to the dismay of her lawful husband, whom the bishop considers to have forfeited both his wife and her valuable dowry. The *Causes célèbres* of Gayot de Pitaval recounts another version of this memorable story at length, the two protagonists being the son and daughter of two merchants living in the Rue St. Honoré in Paris.

A third memorable tale is that of the Lecherous Monk, which can also be found in the *Causes célèbres*. A young French gentleman is forced to become a monk by his religious parents, without having a vocation. While traveling to the monastery, he stops at an inn. The innkeeper persuades him to watch over the corpse of his beautiful young daughter, who died the day before. At the sight of her, the monk "forgot the sanctity of his vows and took liberties with the corpse." After the monk's departure, the apparently dead girl comes back to life. As fate would have it, the monk returns to the inn nine months later and, to his no small surprise, sees the girl alive and with a newborn child in her arms. He at once tells the parents that he is the child's father, casts off his monkish gown and offers to marry their daughter. The innkeeper and his wife are delighted to have this handsome and well-mannered young man as their son-in-law, although the lecherous ex-monk confessed, in so many words, to having raped their daughter while he presumed her to be a corpse.[20]

A FOURTH MEMORABLE story is that of the Careless Anatomist. By far the most famous version of this tale concerns none other than the famous anatomist Andreas Vesalius, author of the *De humani corporis fabrica*. According to a widely disseminated legend, Vesalius was once consulted by a Spanish noblewoman. After the death of his patient, he wanted to discover the origin of her illness and asked the family's permission to open the body. To the horror of both Vesalius and the noblewoman's relatives who were present to watch, the heart of the presumed corpse was found to be still beating. The family accused Vesalius of murder and denounced him to the Spanish Inquisition; had King Philip II of Spain not interceded in his behalf, he would have been sentenced to death. The tribunal granted Vesalius a pardon, but only on the condition that he make a pilgrimage to Jerusalem and Mount Sinai to expiate his crime.[21] It is true that Vesalius left Madrid for Venice in early 1564, but this was because the king had sent him there on a diplomatic mission. After some weeks in Venice, Vesalius went to Jerusalem, for reasons that have been debated. Was he banished as in the tale of the Careless Anatomist, did he go to Jerusalem on a pilgrimage or on a research expedition, or was he simply, as Father Sweertius put it in his *Athenae Belgicae* of 1628, "weary of life at Court and of the brawlings of his wife"? At any rate, Vesalius never returned: on the return journey from Jerusalem, the ship was almost wrecked in a tempest, and Vesalius died on board "in a catarrh" and was buried on the island of Zante, between Cyprus and Greece.

The earliest allusion to the tale of the distinguished anatomist who dissects a patient prematurely comes from a contemporary of Vesalius.

The celebrated French surgeon Ambroise Paré wrote, in his 1571 treatise *De la génération*, that he did not advise that the opening of the dead body be performed in undue haste, particularly when regarding the dire fate of a famous anatomist who was then residing in Spain. The anatomist had opened the body of a woman thought to have died from "suffocation of the womb," but at the second stroke of the anatomist's razor, she began to move and give other signs that she was still alive. The famous anatomist was exiled from Spain and died of grief shortly afterward. Paré does not name Vesalius as the Careless Anatomist, but Fortunio Fideli did so in his 1602 treatise *De relationibus medicorum*. The full version, with details about the Inquisition and the intercession of the king, was published in Melchior Adam's *Vitae germanorum medicorum* in 1620 and was alleged to have derived from a letter written by the diplomat Hubert Languet in 1565. Already in the nineteenth century, a learned biographer of Vesalius pointed out that no person who had known Vesalius during life mentioned his fatal error, not even his friend Delecluse, who came to Madrid just when Vesalius left this city.[22] Nor was the error of Vesalius at all mentioned in the annals of the Spanish Inquisition.

The Vesalius legend is likely to have been influenced by another tale concerning an illustrious Spaniard, Cardinal Diego di Espinosa, bishop of Siguenza and prime minister to the aforementioned Philip II of Spain.[23] One day, King Philip spoke harshly to this gentleman and reminded him "that he was speaking to the president of Castile." The cardinal, who until that moment had himself been president of Castile, understood that the king had dismissed him ignominiously from this office, and fell to the ground as if stunned. He had no pulse and did not breathe; every person present marveled that the king's wrath had slain his minister. It was then decided that the cardinal should be embalmed. But when the dissection was well advanced, the cardinal awoke with a scream of agony and attempted to struggle with the Careless Anatomist. All attempts to save him were of no avail, and the cardinal died before the comforts of religion could be administered to him.

III

Winslow the Anatomist and Bruhier the Horror Monger

To-day is a thought, a fear is tomorrow,
And yesterday is our sin and our sorrow;
And life is death
Where the body's the tomb
And the pale sweet breath
Is buried alive in its hideous gloom

—*T. Beddoes,* A Dirge

The anatomist Jacob Winsløw was born in 1669, the eldest of thirteen children of Peder Jacobsen Winsløw, dean of the Protestant Church of Our Lady in Odense, Denmark.[1] Winsløw entered the University of Copenhagen in 1687 to study theology, but after four years switched to anatomy and surgery. He enjoyed no success as a surgeon, however: at one of his first attempts to perform an operation, the sight of the flowing blood alarmed him so much that he determined never again to try any surgical procedure. Such a state of affairs would, today, have led to his hasty exodus from medical school, but the seventeenth-century Danish professors were more lenient, and perhaps also wiser. They advised him to concentrate on the study of anatomy, and fortunately young Winsløw had no qualms about cutting up the dead. As the prosector of the able anatomist Caspar Bartholin, he made such swift progress that he was awarded a royal grant to travel to Holland and France for further studies. When, in 1699, Winsløw was studying anatomy in Paris, he was deeply influenced by the Catholic tracts of Jacques-Bénigne Bossuet. His spiritual conversion was complete after he met and spoke with Bossuet: Winsløw converted to Roman Catholicism and changed his name to Jacques-Bénigne Winslow.

Winslow was unwilling to return to Protestant Denmark as a Catholic, and Bossuet and his other French Catholics benefactors helped him resume his work in Paris. The medical faculty of Copenhagen was one of the finest in Europe in the late seventeenth century, and Winslow had received a first-rate medical education. It is no coinci-

A line engraving of Jacques-Bénigne Winslow, by C. N. Cochin Jr. after A. L. Romanet. *From the author's collection.*

dence that he rose rapidly in the Parisian medical world. He became physician to several hospitals, a member of the Académie Royale des Sciences, and *docteur-régent* of the Paris Faculty of Medicine in 1728. Winslow was also a distinguished anatomist, who in 1732 published *Exposition anatomique de la structure du corps humain*, a worthy treatise on descriptive anatomy that remained in use well into the nineteenth century. The foramen Winslowii, the opening between the greater and lesser sacs of the peritoneum, still bears his name. Some contemporaries were of the opinion that in spite of his superior knowledge of anatomy, Winslow remained a credulous, superstitious man.[2] This judgment may well have been based on religious bigotry, however, and is not supported by a study of Winslow's published works, which often demonstrate both scientific reasoning and a firm critical judgment. In 1743, at the age of seventy-four, Winslow was belatedly promoted to full professor of anatomy at the Jardin du Roi. He held on to this prestigious post until 1758, when extreme deafness forced him to retire, at the age of eighty-nine. The hardy old Dane died in 1760, without ever seeing his native country again.

Jacques-Bénigne Winslow wrote his anatomical textbook and his articles in the *Memoires* of the Academy of Sciences in French, but all his formal theses at the Faculty of Medicine were in Latin. His thesis *Morte incertae signa*, published in 1740, was one of these formal academic publications and would have been given little attention, had its subject and conclusions not been so startling: Winslow postulated that the signs of death used by the medical profession were unreliable and that people were at immediate risk of being buried alive. The thesis begins with the solemn words "Death is certain, since it is inevitable, but also uncertain, since its diagnosis is sometimes fallible."[3]

JACQUES-BÉNIGNE WINSLOW boldly claimed, "It is evident from Experience, that many apparently dead, have afterwards proved themselves alive by rising from their shrowds, their coffins, and even from their graves." He quoted the works of Zacchia and Lancisi discussed in the preceding chapter, and cited the sad cases of the emperor Zeno, who was willfully locked into a vault, and John Duns Scotus, who was buried alive. Winslow himself added some more recent observations: a certain Mme Landry had told him that her father had once been laid out for dead but had revived when a woman poured salt water into his mouth, and a clergyman named Joseph Mareschal attested that in 1714 he had seen a woman sit up in her (fortunately open) coffin, when the pallbearers passed through Rue Jean Robert in Paris. A more sinister observation was communicated to Winslow by the Paris surgeon M. Bernard. He had been present when the "corpse" of a monk buried three or four days earlier was brought out from a church vault, after a noise had been heard. The monk was "breathing and alive, with his Arms lacerated near the Swathes employed in securing them." M. Bernard was a respectable medical practitioner, and there is little reason to doubt his veracity.[4]

A phenomenon that seems to have worried Winslow even more was the risk that an anatomist might, through a fatal mistake, start dissecting the body of a person falsely diagnosed as being dead, as the great Vesalius was thought to have done. Alerted too late by the still sentient victim's "mournful Shrieks and Cries" upon feeling the pain of the incisions, the anatomist could do nothing but watch the person die by his own handiwork; "that lamentable Circumstance exposed the unwary Operator to eternal Infamy, and the implacable Indignation of the surviving Friends." The Paris surgeon Philippe Peu had once performed a cesarean section on the corpse of a woman, who turned out to have a spark of life still in her; she had trembled and ground her teeth upon the touch of the scalpel, and this blunder had filled the surgeon's mind with terror.[5] It is difficult not to link Jacques-Bénigne Winslow's horror that one of his cold, insensate subjects in the dissection hall might one day give a scream of pain when the first, deep incision was made, the warm, pulsating blood flowing over the hands of the unwary operator, with the known fact that already as a medical student he had a great fear of blood and was incapable of performing surgical operations.

That Philippe Peu had died already in 1706 and that all of Winslow's contemporary cases date from before 1714 suggest that he had written his dissertation at about this time, but refrained from publishing it.[6] It certainly had a much more personal tone than the run-of-the-mill medical thesis of the time. Winslow passionately warned his colleagues about the dangers of apparent death, premature burial, and premature dissection and boldly asserted that "vast Numbers of Persons, who have been too soon interr'd, probably call aloud from their Graves for a due Vengeance on those who have barbarously exposed them to a violent Death." He even claimed that he himself had twice, once as a child and once as an adolescent, been mistakenly laid out for dead and consigned to be buried during his years in Denmark, but fortunately he had revived on both occasions. A later Danish writer made a determined effort to find independent evidence for these occurrences, but without success, although Winslow had written, in a letter to his nephew, that his health had been very precarious from his childhood until he was fifteen years old.[7]

The main point of Winslow's thesis was that although the modern, surgical tests of death or life were better than the primitive, traditional signs of death used among the people, they were still too uncertain to be relied upon. The onset of putrefaction was the only reliable indication that an individual had died. Winslow did not consider the lack of respiratory movement and the absence of arterial pulsations to be infallible signs of death; only putrefaction and the appearance of "livid spots" were sure signs that the individual had really expired. He advocated that no lifeless patient, who could not be safely diagnosed as dead, be shrouded and put into a coffin. Instead, the presumed corpse was to remain in a warm bed and there be vigorously resuscitated.[8] The individual's nostrils were to be irritated by introducing "sternutaries, errhines, juices of onions, garlic and horse-radish." They could also be tickled with the quill of a pen, while some preferred thrusting a sharp, pointed pencil up the corpse's nose. The gums were to be rubbed with garlic, and the skin stimulated by the liberal application of "whips and nettles." The intestines could be irritated by the most acrid enemas, the limbs agitated through violent pulling, and the ears shocked "by hideous Shrieks and excessive Noises." Vinegar and pepper should be poured into the corpse's mouth "and where they cannot be had, it is customary to pour warm Urine into it, which has been observed to produce happy Effects." The corpse that withstood these brutal methods of resuscitation faced even more unpleasant experiences ahead. The soles of the feet should be cut with razors and long needles thrust under the toenails. Whereas Lancisi recommended that a red-hot iron be applied to the soles of the feet, Winslow preferred pouring boiling Spanish wax on the corpse's forehead; a French clergyman suggested that a red-hot

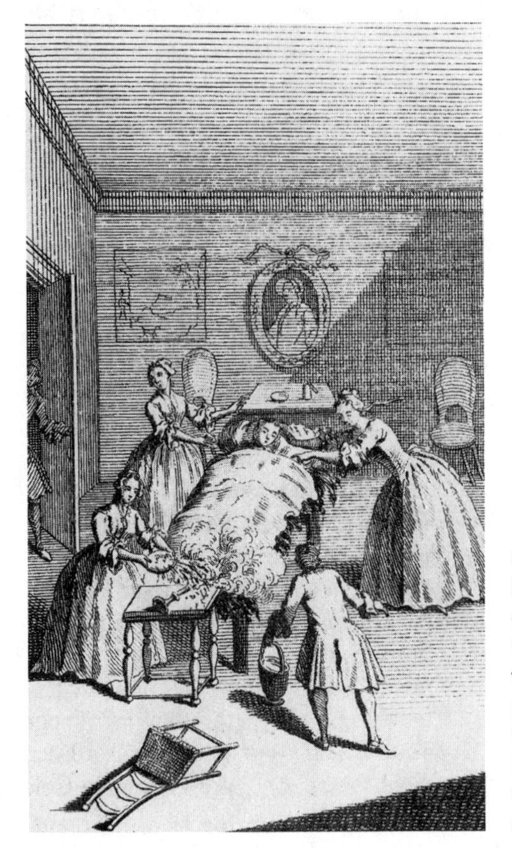

An apparently dead lady, who has previously withstood several determined attempts of resuscitation, awakes with a shriek when some careless young people set her bed on fire by mistake. Another illustration in the 1746 English edition of Bruhier's book. *Copyright British Library.*

poker be thrust up the unfortunate corpse's rear quarters as a last resort. Although these violent and barbarian measures seem ludicrous today, and more worthy of the torture chambers of the Marquis de Sade than the mortuary of a French hospital, Jacques-Bénigne Winslow was guided solely by his humanitarian feelings.

IN SPITE OF its startling contents, Winslow's thesis on the signs of death would have seemed destined first for a short shelf life in the hands of a few academic booksellers and then for an eternal rest in the sepulchral vaults of selected European university libraries. The quality of these formal academic theses was variable, the editions generally small, and the readership limited to a few medical scientists, most of them within the Paris faculty. But Winslow's thesis caught the eye of the physician and translator Jean-Jacques Bruhier d'Ablaincourt. Born in Beauvais in 1685 and educated at the University of Angers, Bruhier spent most of his life as a reputable and busy medical practitioner in Paris.[9] Like Winslow, he was a member of the Academy of Sciences, and took active part in its transactions. But unlike the Danish anatomist, he was quite unknown in European medical science; indeed, he had not written a single original medical treatise in his life. Bruhier translated a variety of German and Latin books, however, some of them useful medical and obstetrical textbooks, others repositories of amusing anecdotes. These translations show him not only as a skillful linguist, and a scholar with considerable understanding of European history and culture, but also as an author with a readable style and—more unusual, by mid-eighteenth-century standards—an eye for what the reading public wanted. It is likely that Winslow's thesis on the signs of death had caused some stir when it was discussed by the Paris faculty, and Bruhier must have realized that he had come on to a good thing. He visited a meeting at the Faculty of Medicine and congratulated Winslow on his thesis; the elderly Dane replied that he had in fact himself tried to have it translated into French, but been displeased with the result. The wily Bruhier then introduced himself as a literary man and offered to prepare a superior translation.[10] Winslow, who probably knew about Bruhier's earlier work as a translator, readily accepted, little knowing what far-reaching consequences this decision would have in European culture.

Although there is no earlier record of Bruhier's being in any way concerned about the risk of apparent death and premature burial, in his edition of Winslow's thesis he emerges as a fully fledged campaigner for burial reform. Bruhier declares himself convinced that putrefaction is the only certain sign of death, that the prevailing routines for declaring people dead are lamentably fallible, and that his fellow Frenchmen are in immediate danger of being buried alive. In a lengthy addition to Winslow's thesis, he elegantly formulates his ambition: "As an author, therefore, who is desirous of being universally useful . . . I shall add to the Histories collected by Mr. Winslow, in order to prove the Uncertainty of the Signs of Death, some others, whose Multiplicity will justify the Precautions of the Prudent, destroy the vain Pretexts of the Incredulous, make deeper Impressions of the Minds of the Giddy, and alarm those whose Insensibility calls for the most powerful and commanding Evidence."[11] To bolster the contents of this slender volume, Bruhier also contributed a treatise on the amusing and exotic funerary customs of various tribes around the world, which was in part extracted from one of his earlier translations.

Bruhier quoted the miracle of the reviving Cardinal Andreas from Kornmann's *De miraculis mortuorum*, but scoffed at the credulous historian's interpretation of this as a divine intermission. Kornmann and Garmann had also been at a loss to explain how some dead people could chew and swallow their shrouds in their graves, or how a certain dead woman had begun to devour her own body. These were no miracles or cases of vampirism, Bruhier wrote, but "shocking Phenomena bred from a despair natural to a person buried alive." Bruhier had found some more credible accounts in a letter from a certain Dr. J. J. Crafft to the chronicler Simon Goulart, which presented several cases of fatal or near-fatal mistakes in diagnosing death. One of them concerned Jean de

Lavaur, a nobleman of Neuchâtel, who was brought back to life in his coffin when his physician blew powdered pepper into his nostrils; another, a young lady in Augsburg who had been buried in a vault and whose body, with no fingers on the right hand, was found on the stairs of the vault when it was opened some years later. About the year 1730, a poor woman in St. Germains had been laid out for dead. Some young people were supposed to wake over her, but they instead diverted themselves with various games and accidentally turned over a wax candle at the foot of her bed. The bedclothes caught fire, and the woman gave a hideous shriek.[12] When recounting a tale of a man who had revived in his coffin when a friend sprinkled his face with holy water, the scoffing Bruhier again wanted to "apprize the Reader, that this Effect was produced by the *Coldness*, and not the *Consecration* of the Water, whose genuine Qualities can never be altered by a Rhapsody of superstitious words." He even had the audacity to comment that Jesus Christ would never have been capable of raising Lazarus from the dead if the onset of putrefaction had been awaited before the man was declared dead![13] It is apparent from this rationalist, sneering attitude toward religious ritual, which is frequently expressed in Bruhier's books, that he himself was not a practicing Catholic. In his view, the miracles of the resurrected dead in the works of Kornmann and Garmann were nothing of the sort, just distressing instances of premature burial.

To prove that a person might live for quite a long time in the absence of both heartbeat and respiration, Bruhier quoted some astounding case reports from William Derham's *Physico-theology*.[14] The original source was a 1676 thesis written by Professor Johann Nicolas Pechlin, a Swede who had been active at the University of Kiel, in Germany. Pechlin had spent much time searching for instances of people being able to survive for considerable periods of time without food or without access to air. Like quite a few other medical men, he was much impressed with the various current stories about "fasting girls" who claimed to subsist for months, or even years, without taking nourishment of any kind. His most astounding case of a human being surviving underwater was that of Erik Björnsson, a gardener at Drottningholm Castle, outside Stockholm. In the winter of 1646, the gardener had tried to save a drowning man. He valiantly leaped into the ice-cold water, but himself got stuck under the thick ice. He survived underwater for sixteen hours, before being saved by a boatman who thrust a hook into his head with great force and pulled him up alive. The submarine gardener was received in audience by Queen Hedvig Eleonora of Sweden, who examined the mark of the hook in his head and gave him a pension for life. Björnsson said that the intense cold had made his body completely stiff and insensible, but that the water had been prevented from entering his mouth and nose by a bubble of air. The only sound he had been able to hear under the ice was that of the church bells of Stockholm. This amazing case is well verified by several contemporary Swedish writers, who had all met the submarine gardener and heard him tell his story. Some of Pechlin's other stories were less trustworthy, however: a Swedish woman named Margaretha Larsdotter claimed to have subsisted three days underwater, and a certain Laurens Jonas, a Laplander living near the town of Piteå, was adamant that he had once "drowned" for six weeks before being rescued. Bruhier was not averse to adding some other old idle tales: a German lady had lived three days under the waterline, and a certain Gocellinus, the nephew of the archbishop of Cologne, had taken a fifteen-day dive before emerging alive and well. After reviewing these astounding examples of the extreme capacity of the human body to survive without air, Bruhier declared that both the absence of a heartbeat and the absence of respiration were worthless as signs of death.

While admitting that Winslow's regimen of cutting, pinching, and burning the corpse to make sure that life was extinct had its merits, Bruhier suggested a far more extensive reform of the current burial practice.[15] There should be a system of morgues, or receiving houses for the dead, staffed with competent physicians and watchmen and financed by the state. Here, each corpse should be supervised for at least seventy-two hours, or until putrefaction had set in. It is typical of the rationalist, anticlerical sentiments of Bruhier that he actually suggested that priests be excluded from tendering any services in his model mortuaries. In 1745, he presented this proposal to King Louis XV, and was courteously received in audience by him. His Majesty was apparently much impressed by the Paris physician's gruesome lecture, since he half promised that the state would finance the establishment of a corps of mortuary attendants, with the express purpose of saving those mistakenly thought dead. But when the king's ministers considered the costs of this reform, they persuaded Louis XV to postpone it indefinitely, to Bruhier's chagrin.

JEAN-JACQUES BRUHIER'S book was an immediate success: it was widely read and kept its popularity for many years. Bruhier himself took an active part in promoting the book, submitting extracts to newspaper editors and sending free copies to all medical academies in France. To widen its appeal abroad, he approached the French ambassadors in various countries, and also the foreign representatives at the court of Louis XV at Versailles. The reviews in leading periodicals like the *Journal des Sçavans* and the *Mercure de France* were excellent, and the academies sent glowing certificates of approval to the proud author in Paris.[16] By the age of fifty-seven, Bruhier had finally become famous; he would remain devoted to the cause of burial reform for the remainder of his life. The only critical voice to be heard was that of the abbé Desfontaines, a libertine and controversialist, whose major claim to fame was that he had once been a protégé of Voltaire's, before quarreling with his famous master.[17] In a hostile review of Bruhier's book, the abbé claimed that many of the doctor's stories of swooning ladies falsely declared dead belonged to the realm of fiction and legend. This was a credulity unbecoming to someone masquerading as a scholar and writing on a subject of national importance. *He* was not a medical practitioner, the abbé drily added, but it seemed quite *hazardous* to postulate, as a matter of physiological fact, that a person could live for several weeks underwater.

As could be expected, Bruhier took great exception to this acerbic attack. The abbé Desfontaines had a poisonous pen and knew where to plant a barb where it hurt the most, and the accusation that he had been both credulous and lacking in medical judgment hurt Bruhier's amour propre, which had been considerably padded by the many laudatory reviews and accolades he had received. In 1746, he published what was purported to be a second volume of the *Dissertation*. This 540-page book had three aims: to thoroughly dispose of the objections of the abbé Desfontaines, to introduce a number of new cases of premature burial and lucky escapes that had been sent to Bruhier from readers all over France, and to restate his main arguments about burial reform. Jacques-Bénigne Winslow's name was not on the title page of this second volume, and his original thesis was hardly even mentioned; Bruhier now felt safe as the leader of the campaign for burial reform and preferred to leave the elderly Dane behind.

In his second volume, Bruhier scoffed at the abbé's arguments that Bruhier's view of death as a process rendered it unclear when the soul departed the body and that his theory of death and resuscitation challenged the miracle of resurrection. Bruhier pompously declared that, as a rationalist physician, he rejected superstition of this kind and did not allow it to affect his work for the public good. Bruhier was skating on extremely thin ice, however, when discussing the abbé's reservations about some of his ancient case reports. Instead of admitting that some of them were poorly authenticated, but that it was of much greater importance that he could present such a multitude of modern instances, he stubbornly defended every one of his cases, sometimes with ludicrous arguments. Bruhier devoted thirty pages to gather the evidence in the case of Cardinal Andreas, quoting extensively from the annals of the saints. His defense of the old tales of submarine Swedes was similarly feeble: he claimed that since Pechlin's stories had been vouched for by none less than the famous Danish anatomist Thomas Bartholin, it was unbecoming for a layman like the abbé Desfontaines to doubt their truth. Bruhier vigorously marketed the new volume in his usual manner: liberal quantities of free copies were sent to kings, academies, and ambassadors. In the Royal Academy of Angers, the poet Claude-François de la Sorinière read a laudatory poem to the famous author and propagandist for burial reform. He envisioned that after the publication of Bruhier's two volumes, valetudinarians in every part of France were busy rewriting their wills to prevent their greedy heirs from having them put into the earth too hastily, and to warn their elderly and shortsighted family physicians to take care before declaring them dead.[19] M. de la Sorinière also provided Bruhier with a curious anecdote. The keeper of his vineyard, named Gazeau, had once heard a knocking sound coming from inside a coffin when serving as pallbearer at a funeral in Angers. But M. Gazeau, a timid, religious man, did not tell anyone about what he had heard, since he was fearful of disturbing the funeral ceremony. "Voilà des délicatesses d'un goût singulier"—"See here a singular instance of modesty"—was Bruhier's comment.[20]

In 1749, the ever-diligent Bruhier had another project ready. He had revised the first volume of the *Dissertation*, making each of Winslow's paragraphs into a separate chapter, omitting the text of Winslow's latin thesis, and not even mentioning the Dane's name on the title page. Bruhier was by far the more ambitious and vigorous of the two, and he had by this time completely ousted his original coauthor. Bruhier's two-volume *Dissertation sur l'incertitude des signes de la mort* is nothing if not impressive: it would take a very audacious and imprudent reader to shrug it off and to claim that the risk of premature burial in early eighteenth-century Europe was negligible. The tomes were well written and convincingly argued, although the number of cases—181 in all—made them a little repetitious. It must have been evident to the reader that Bruhier had wide knowledge of the older literature on the subject and considerable experience as a physician.

IT IS APPARENT that Bruhier was much influenced by the memorable old tales of the Lady with the Ring and other prematurely buried heroines. He immediately fastened on the Richmodis von Aducht story, which he had read in the travel book of Maximilien Misson, as a particularly old and curious instance of rescue from the tomb. He may well have visited Cologne to see her grave "under a lofty and magnificent Monument of Stone. In order to perpetuate the Memory of her Fate, there was affixed to the Monument a large Piece of Painting, in which the Accident was described in picture & verse."[21] What follows is more flattering to his determined research and his love of a good story than to his diligence in checking his facts. In the 1742 edition of the *Dissertation de l'incertitude des signes de la mort,* he presented two other versions of the Lady with the Ring story, from Toulouse and Poitiers. In the extended version of his book, he offered no fewer than seven different versions of this pan-European legend, adding new cases from Leipzig, Dublin, and Bordeaux, without any indication of doubt-

An apparently dead lady awakes from her trance when a thief attempts to steal a valuable ring on her finger. Another illustration in the 1746 English edition of Bruhier's book. *Copyright British Library.*

ing the truth of any of them, or suspecting that they might have a common origin.[22] When preparing his second volume in 1746, Bruhier did his best to defend the veracity of the Reichmuth von Aducht story against the onslaught of the abbé Desfontaines: he claimed that the high opinion of the legend in old Cologne and its having once been quoted in a medical treatise by the learned Velschius were strong arguments in its favor. In addition, he repeated three different versions of the tale of the two young lovers in his *Dissertation de l'incertitude des signes de la mort*: the one from the *Causes célèbres*, one that a friend in Bordeaux had told him concerning a dashing young officer in Périgord, and one from Languedoc.[23]

Thus there is good evidence that the abbé Desfontaines was right: quite a few of the stories in Bruhier's *Dissertation* were derived from popular legends and folktales, which had been made into medical case reports by way of various repositories of amusing stories and *Causes célèbres*. Was Bruhier himself aware of this? He might have been, since he used the argument that an *unproven* fact was better than a *false* one when defending himself against Desfontaines in 1746. In 1749, after quoting the tale of the Lecherous Monk in full from the *Causes célèbres*, he added that M. de Pitaval probably had fewer demands for historical proof than Bruhier himself wanted for his *authentic* cases.[24] Bruhier was an author with a good grasp of what people wanted to read, and the "amusing" stories of ladies saved from the tomb served him well as a contrast to the darker tales of gnawed fingers and scratched coffins.

But what are we then to make of the remainder of Bruhier's case reports: Are they at all reliable, or was there such a wealth of popular legends and stories about fatal or near-fatal mistakes in declaring people dead? The persona of the apparently dead *woman*: the sleeping beauty, impossible to awaken from her trance, was a subject of gruesome fascination. The attempts to resuscitate her motionless, limp body had a perverted sexual appeal akin to necrophilia, and the torments of the prematurely buried woman awakening in the tomb, as described by

Bruhier and other writers, had strong sadistic overtones. A wide variety of popular legends featured more or less unwholesome variations on these themes. When researching the 1742 edition of his book, Bruhier had trawled the European literature for interesting cases; in the 1746 and 1749 additions, he also relied heavily on cases sent to him by colleagues and lay readers of the original volume. This exposed him to many folk-tales of doubtful veracity. Bruhier's cases often expressly state that the heroine not only survived the incident for many years but was in full health, completely cured of the disease that had brought her to the brink of the grave, and had numerous children. This strikes a modern reader as suspect. A ludicrous example is provided by the story of Mme Rousseau, wife of a merchant in Rouen, who fell into a deathlike stupor. Her doctor thought her dead, but M. Rousseau persuaded him to try every method to revive her: her body was covered with many leeches, and the soles of her feet were cut twenty-five times with a razor. After the twenty-sixth cut, which was deeper than the others, she woke up and said, "Ah, que vous me faites de mal!"—"Ah, how you are hurting me!" She then lived for many years in the prime of health, and had twenty-six children, one for every cut![25]

Another type of story has the motif that Death must have his prey, although the victim might be changed. This is exemplified by the grave robber who drops dead when the Lady with the Ring awakens; another version has the devoted husband or heartbroken mother fall dead from joy and surprise when their apparently dead wife or child is brought before them. Nor can we believe ludicrous stories like the one about the Count Richard, who went into a church late at night to say his prayers. In the church was an open coffin, and in the coffin a presumed corpse. Suddenly, the dead man stood up and advanced toward the count, his hands outstretched to embrace him. Count Richard believed in ghosts, however, and ran him through with his sword. Thus Death had his victim after all. The ending is that the remorseful Count Richard issued a proclamation throughout his area of jurisdiction that it was wise to make sure that people were really dead before they were buried.[26]

It must be admitted that these fanciful additions detract from his case, but Bruhier aimed to amuse as well as to instruct, and to make his book accessible to the educated layperson. In fact, many of his cases were contemporary. Of these, a great majority involved narrow escapes, where the presumed corpses either had made their resurrections known in their beds or their coffins or had been saved by sharp-eyed medical attendants. These reports tend to be brief and matter-of-fact; they add impressive evidence that such incidents were by no means rare. Many of Bruhier's cases were attested by the local doctors, vicars, or schoolmasters. One from Dole was confirmed by M. Charles, professor of medicine in Besançon. A troop of soldiers had been allowed to make camp in the local cemetery. While some of them were strolling about among the graves, they heard a faint cry coming from one of the vaults. These soldiers, not afraid of ghosts, promptly broke down the vault door and rescued a young girl who had been interred a few hours earlier. It turned out that the girl had been gravely ill for some time and that her *maîtresse* had presumed her to have died. Too parsimonious to call a doctor, she unceremoniously had the coffin dumped in the family vault. Another sinister example came from M. Rigadoux, a surgeon and obstetrician from Douai. In September 1745, he had been called to attend the wife of François Dumont in the village of Lowarde. But this village was quite far away, and when M. Rigadoux arrived, the husband said that his wife had died two hours earlier. She had been put in a coffin, and the surgeon agreed that she showed every sign of death: there was no pulse, no heartbeat, and no respiration. Her belly was grossly distended, and the husband wanted M. Rigadoux to perform a cesarean section. The dexterous surgeon instead managed to turn the child and deliver it by the normal route. The child, too, appeared motionless and dead, although M. Rigadoux did his best to revive it with warm blankets and by rubbing it with hot wine. He then went to dinner with the curé, leaving the child in the hands of the local midwives. When he came back to check on his patient, he was delighted to find the child had made its resuscitation known in a strong voice. It then occurred to him that the mother might not be dead either: she was taken out of the coffin, put in a warm bed, and rubbed with wine. Sure enough: she gave faint signs of life after several minutes of treatment, and when M. Rigadoux revisited the family some weeks later, she was working in the household, although the incident had left her lame, deaf, and feebleminded.[27]

Had Bruhier chosen to present only a dozen or so of the contemporary cases he had collected, the result would have been a scholarly paper that made a convincing case for the fallibility of the current application of the signs of death, while lacking the accessibility to a wider audience. Such a paper would probably have been quickly forgotten. The medical establishment was unwilling to admit that the signs of death were uncertain, and the clerical establishment was unlikely to be impressed by Bruhier's anti-Christian gibes, or by his plan to exclude priests from his model mortuaries. By presenting his case directly to the educated public, Bruhier bypassed these opponents, but to do so successfully, his books had to be amusing and readable. The tales of the Lady with the Ring and of the Lecherous Monk, and their numerous variations, had considerable staying power in European folklore and did much to spice up Bruhier's books, and to find them new, eager readers.

BRUHIER'S BOOK WAS spread through large parts of Europe. In 1744, the first volume of the *Dissertation sur l'incertitude des signes de la mort* was translated into Italian. The year after, the celebrated physician Richard Mead published the third edition of his *Mechanical Account of Poisons* and mentioned Bruhier's book briefly. A reviewer in the *Gentleman's Magazine* found this reference particularly interesting and gave a separate account of it, fully agreeing with Bruhier's doubts about the currently used signs of death.[28] Not long thereafter, an English translation of Bruhier's book was published. Entitled *The Uncertainty of the Signs of Death*, it lacked the author's name on the flyleaf; it might well have been an unauthorized translation, aiming to cheat the French authors on the profits from this best-seller.[29] The book was a verbatim translation of Bruhier's 1742 edition, with the addition of a garbled account of the case of the lamented Madam Blunden, of Basingstoke. In the book, she becomes a noblewoman of character and fortune, who is buried in the family vault; the translator further improves the already gruesome story by stating that she was actually alive and conscious when rescued by the sexton, although the injuries she had inflicted on herself in the tomb were such that she died in a few hours of inexpressible torment. Unlike the French original, the 1746 London edition was embellished with six plates, illustrating the dramatic stories of the Lady with the Ring and some other prematurely buried heroines. This edition was reprinted in Dublin in 1748, without the six plates. In 1751, a second edition was printed for Mr. M. Cooper, at the Globe in Paternoster-Row. Its preface, which the bookseller Cooper had probably written himself, stated that it was the duty of all Englishmen to read it and become acquainted with the fallibility of the signs of death currently used; if they did not, they would have blood on their own hands when their apparently dead wives or children were carried away and "placed in so dreadful a Situation, cramped up in the close and dark Confinement, and forced by Nature to struggle in vain for Life."[30]

The third foreign translation of Bruhier's *Dissertation sur l'incertitude des signes de la mort* appeared in 1751. Dean Tillaeus, rector of the French Lutheran church in Stockholm, had read and been horrified by the French original. After a wealthy merchant agreed to pay the costs for this philanthropic endeavor, Tillaeus translated the 1742 edition of Bruhier's book, himself adding a postscript about some curious Swedish instances of fatal mistakes at the end of life.[31] One of these was considered particularly amusing. In 1664, the elderly Lieutenant Färling died, it was presumed, and his body was shrouded and put in a coffin. The morning of the funeral, a female servant carried some firewood into the room with the coffin. She was startled by a voice behind her, ordering that she bring his trousers and slippers. When she turned around, the deceased lieutenant was sitting up in his coffin. She fainted dead away, and when a servant led the old man into a room where a crowd of gravediggers and servants were eating their breakfast, a wild stampede broke out among the superstitious Swedes, who were fearful of ghosts. Several of them fell over in their haste to get out of the room, and a woman was disfigured for life with a pugilist's nose. The dean found it very gratifying that the resurrected lieutenant, formerly a drunkard and wife beater, became a devout Christian, who spent much of his time reading the Bible. Dean Tillaeus also knew a more sinister instance of apparent death, which had been reported by a colleague of his, Dr. Carl Johan Lohman, rector of Tierp. In 1706 or 1707, a drowned mariner was buried at a churchyard in Stockholm. A dull thud was heard as the earth was poured down onto the coffin, and the rector was called. He ordered the coffin to be exhumed, and the many people who congregated around the coffin as it was opened saw that the mariner had kicked out the bottom of the coffin and that his shroud was drenched with fresh, warm blood, which smoked in the cold air. This sign, and others too horrible to mention, indicated that the man had been buried alive and died in his coffin. A later Swedish investigator verified both these instances from independent sources.[32]

The fourth, and most important, translation of Bruhier's *Dissertation sur l'incertitude des signes de la mort* appeared in 1754. A wealthy Danish lady had paid Dr. Johann Gottfried Jancke, teacher of medicine at the University of Leipzig, to translate the entire text of Bruhier's two volumes into German.[33] In his preface, Jancke, who knew and admired Jacques-Bénigne Winslow, rightly questioned Bruhier's motives in completely excluding the elderly Dane's name from the title pages of his 1746 and 1749 editions. Had not Winslow's thesis inspired the entire project, and could Bruhier really stand up and say that the book that had made him famous all over Europe was entirely his own work? Except for this objection, Jancke was a wholehearted admirer of Bruhier's concept of putrefaction as the only certain sign of death. The publication of the German translation of Bruhier's book was to have profound effects: it changed the concept of death even more effectively in Germany than in its native France and was eagerly adopted both by the medical profession and by the general public. There was much debate about the uncertainty of the signs of death and about the best way to save people from being buried alive. From 1754 until well past the end of the century, a veritable torrent of serious theses, alarmist pamphlets, and amusing books of anecdotes about swooning ladies rescued from the tomb were published in the German language, all of them deriving a good deal of their subject matter from Jancke's translation of Bruhier.

IV

The Eighteenth-Century Debate

Parted from God the soul is dead
Buried alive the graceless soul
His conscience as with worms o'erspread
No sepulchre is half so foul!
—*C. and J. Wesley,* Short Hymns on Select Passages of the Holy Scriptures *(Bristol, 1762), hymn no. 1975*

Not until 1752 was there any serious opposition to Bruhier's *Dissertation sur l'incertitude des signes de la mort.* Antoine Louis was an opinionated young surgeon at the Salpêtrière hospital in Paris, and a protégé of the celebrated François de La Peyronie, the premier surgeon to the king.[1] He held Bruhier responsible for the premature burial hysteria that reigned in France, by having reproduced as medical facts, in his ill-researched book, a lot of ridiculous stories invented to amuse women and children. The main motive of Louis's attack, it has been presumed, was defending the honor of the medical profession. At this time, the French countryside was full of quacks and itinerant charlatans, who posed a serious threat to the professional medical men. The French peasants were, not unreasonably, of the opinion that if the doctor could not tell a living man from a dead one, he was good for nothing. Bruhier's ideas about the uncertainty of the signs of death thus threatened the authority of the medical profession at large. But there is evidence that Louis was prompted by more ignoble motives. It is very likely that he had observed, with a mixture of jealousy and dismay, the meteoric rise of Jean-Jacques Bruhier within the Parisian medical community: in 1742, Bruhier had been an unknown, middle-aged practitioner; now he was an author of international acclaim, acknowledged as one of the leading medical theorists of the time. One eighteenth-century writer says that Louis was dismayed not only by the spreading fear of premature burial induced by Bruhier's books but also *by their rapid sale*; another, that he felt personal hatred toward Bruhier and envied his fame as philanthropist.[2] The German translator of Bruhier, Professor Jancke, writing two years after Louis's book had been published, dismissed the book lightly and claimed that it was well known that there was a personal animosity between Louis and Bruhier.[3]

Antoine Louis was just twenty-nine years old when his *Lettres sur la certitude des signes de la mort* was published. He had structured it as a series of letters directed to a fictitious correspondent, one of the many honest Frenchmen frightened by Bruhier's *Dissertation.* He declared himself both amazed and dismayed that Bruhier's book had been approved and honored by so many academies all over the country. Louis was a young surgeon, not a historical scholar, and could not add to the arguments of the abbé Desfontaines about the veracity of some of the older cases. The wily Bruhier had used his second volume to prove the dubious tales in the first, with the claim that an unproven fact was better than a *false* one. But it was still wrong, Louis replied, to make vulgar people believe extraordinary things without sufficient proof. Far from finding the tale of the Lecherous Monk in any way suspect, he quoted it at length, and then tried to reinterpret the conclusion: the Monk had not really believed that the girl was dead, since her face, as the story went, had not been disfigured by the horrors of death. Another ludicrous argument of his was that the error of Vesalius, the veracity of which he doubted as little as Bruhier and Winslow had, should not be used to discredit the medical profession, since it was well known that Vesalius had such a passion for the study of anatomy that it sometimes got the better of him. Nor does the following case report, which he discussed at length, aid Louis's cause. In February 1746, a young country girl died at the Hôtel-Dieu hospital. One of the nuns told Louis's pupils that they could take her corpse to the dissection room for use in the anatomy lessons the next day. But late at night, plaintive sounds and pitiful sighs could be heard coming from the deserted anatomical theater. One of the frightened students ran to fetch Louis, who was horrified to see the girl's body contorted in an unnatural manner inside her shroud; one leg had actually been pushed through the shroud and reached the floor. She was now really dead, and Louis admitted to feeling both horror and dismay when he surveyed this lugubrious scene. Using his usual kind of logic, he argued that it had been an ignorant young medical student who had manhandled the "corpse" into the dissection room, not a proper doctor, and that this occurrence, although tragic, was no proof for the uncertainty of the signs of death.

The polemical attitude of Louis did his cause no good, nor did his attempts to use the reductio ad absurdum approach of the traditional medical thesis. Not content with blasting Bruhier for his credulity and medical charlatanism, he launched an attack on the old Winslow. With rather questionable logic, he claimed that if the elderly anatomist really believed that the signs of death were uncertain, he was a mass murderer, since many of his "subjects" in the dissection room would probably still have been alive, and yet he had cut them up. Louis was much offended by Bruhier's accusation that even a trained medical practitioner was unable to tell a living person from a dead one. In reviewing Bruhier's cases, he found that in many of them no doctor had been called to see the apparently dead person; vulgar people, not well-trained doctors, had made the fatal mistakes. And in the case of the nobleman Jean de Lavaur, had it not been his clever physician who saved his life by blowing strong pepper down his nostrils?

Antoine Louis believed that rigor mortis and certain changes in the cornea of the eyes were even surer signs of death than was putrefaction, which might occur in a festering, gangrenous limb. He also objected to the use of putrefaction as the sole sign of death, since the delayed burial of corpses meant that putrefying bodies were left lying around in the dwelling houses for a considerable time. He felt that this was protecting the interest of the dead to an immoderate degree, while exposing the surviving relatives to the emotional distress of having decaying bodies about; furthermore, the corpses spread deadly miasmata of contagion, which might poison the living people in the household. To try to revive apparently dead people, Louis used a remarkable apparatus, made specially for the purpose, to administer enemas of tobacco smoke to awaken those apparently dead. One of the pipes of this contraption, which rather resembled an oversized set of bagpipes, was thrust up the presumed corpse's rear passage, and another was connected, by way of a powerful bellows, to a large furnace full of tobacco. Louis prided himself that several Dutch and German physicians, anxious to know the latest advances in clinical medicine, had been to France to see these enemas of tobacco smoke being administered to the hapless corpses at the Paris morgue.

It is apparent that Louis thought Bruhier a horror monger, an avaricious medical busybody and dilettante, and a traitor who willfully accused the medical profession of ignorance to further his own reputation as a writer. But Bruhier, still alive and well in 1752, thought Louis an envious young whippersnapper of a surgeon, who insolently tried to usurp some of his own fame. In a furious rebuff, the elderly physician took young M. Louis to task in the *Mercure de France* magazine, saying his book was both badly researched and uncouth, and his accusations against more-senior medical practitioners like himself and Winslow ungentlemanly. The way Louis had advertised his book, by pasting up posters all over Paris, and even in Versailles, was disgraceful to the medical profession at large. Bruhier ended by declaring himself absolutely unconvinced by the arguments in Louis's book: the signs of death discussed in it were all more or less fallible. Louis rather meekly replied that it was the bookseller who had put up all the advertising posters, not himself. He also thought that Bruhier had himself behaved in an ungentlemanly manner: politeness was an essential part of any literary or scholarly debate, and he could detect no trace of it in Bruhier's vituperative missive.[4]

IN THE NINETEENTH and twentieth centuries, some positivist medical historians applauded Antoine Louis for what they called his refutation of Bruhier's errors, and his stalwart defense of the medical profession. His thesis—that the signs of death were reliable and that people had not been at risk of being buried alive—agreed well with what people thought (or at least wanted to think) really was the truth.[5] But both the quality of Louis's book and its influence at the time have been considerably overrated. Bruhier's *Dissertation* was elegantly written, and closely and convincingly argued, and presented its medical points in a way accessible to the educated layperson. Louis was a less accomplished writer and also emerges as a narrow-minded zealot, whose refusal to acknowledge any form of wrongdoing on the part of the medical profession must have evoked skepticism even at the time. Many of his arguments were far from impressive, and he could not deny two mainstays of Bruhier's case: people had been buried alive in all ages, and there were some alarming recent instances of fatal mistakes in declaring people dead. The French reviews of Louis's book were mixed, and in later eighteenth-century writings on apparent death, references to it are generally dismissive. Nor did it appear in any foreign translation. Why, if Louis was so successful in calming the minds of the mid-eighteenth-century people, was a great number of books, pamphlets, and theses on apparent death and premature burial published from 1760 until the end of the century, and well into the next one? There is no doubt that Bruhier was the winner of the mid-eighteenth-century French debate on the uncertainty of the signs of death; the majority of the medical men took his side in the controversy and disregarded the objections of Louis.

Bruhier's theories were doubtless also predominant among laypeople. The mid-eighteenth-century upsurge in fear of premature burial can easily be verified in literature, debate, and public sentiment. An interesting study of Parisian wills demonstrates that whereas only 2 out of 1,000 wills from 1710 to 1725 detailed safeguards against premature burial, no fewer than 13 out of 1,000 wills from 1760 to 1777 did so, and a further 34 wills asked for a delay in burial for unspecified reasons.[6] The underlying reasons for the immediate appeal of Bruhier's book, and the fear of premature burial that ensued all over Europe, have been discussed by several historians. The prevailing hypothesis, supported by Philippe Ariès and the majority of later writers, is that the fear of being buried alive was a by-product of the ongoing process of dechristianization.[7] Many people had begun to question the traditional Christian dogma, and the secular rationalism left an emotional vacuum for those who had to confront the thought of death without the hope of paradise. This led to an increased fear of death. Once death had been deprived of the time-honored assurance of the Christian faith, people became concerned about what happened to their dead—or perhaps not so dead—bodies: If they did not enter a better world in the odor of sanctity, what befell them? The physical torments of a premature burial—the ghoulish minutiae of gnawed hands, bruised heads, and beaten bodies that recur in Bruhier's books—may be taken to represent some kind of secularized hell.

But several other factors—sociological, medical, and personal—acted in favor of Bruhier. Wooden coffins were introduced, Ariès tells us, after the thirteenth century, but at this time coffins themselves were perogatives of the wealthy.[8] Their use for ordinary people became increasingly common with time, but in the sixteenth century, many people were still buried in their shrouds, without coffins. In 1569, a communal coffin was used in one part of London, to transport the corpses of the poor to the grave, and an account from Rye in Kent tells us that by 1580, coffins in this town were reserved for its leading citizens. In Scotland, at the same time, a "death hamper" was re-used as a means of transportation for poor people: it was turned over in the grave to deliver its shrouded cargo into the earth. A similar contraption was employed in Brittany.[9] In the 1603 plague epidemic in London, coffins were rare and expensive: the dead were just covered with a winding-sheet before burial.[10] There is evidence that in some parts of England the use of coffins did not fully take hold until after the Restoration. The use of wooden coffins increased considerably in the seventeenth and early eighteenth centuries, for all classes of people except the destitute poor. Since the deceased individual now had private space in death—a close, impenetrable wooden shell buried under several feet of earth—it is not unnatural that anxious and worried people were susceptible to arguments of what would happen if the buried person was not really dead. It is apparent from the eighteenth-century wills detailing safeguards against apparent death that the people were not frightened of merely being mistaken for dead; it was the specific fear that they might awaken in a coffin that prompted their actions. Some individuals specified autopsy, embalming, or even decapitation, to make sure they would really be dead when buried.

In the early eighteenth century, observations from different fields of medicine and natural history introduced a novel uncertainty concerning the organism's capacity for life, without respiration or heartbeat, and without taking any form of nourishment. A curious and still unexplained zoological mystery, that of the "Toad in the Hole," provided some unexpected arguments. Since the Middle Ages, there had been numerous reports of toads found completely enclosed in blocks of stone. Quarrymen and masons were no strangers to these mysterious toads, which were usually still living when freed from their narrow cells in the stone. The original observers of these toads in the hole had viewed them as the products of witchcraft and sorcery; in the late seventeenth century, however, men of science became more aware of this obscure mystery. Had the toads been encased in stone ever since the stones were formed after the Deluge, and how could they live for perhaps thousands of years without access to food, air, or water? The French medical scientist Claude Nicolas Le Cat was not the only person to draw parallels between the slumbering half-lives of the entombed toads and similar trance-like states in human beings.[11] Another of the old zoological myths, that the swallows hibernated on the bottom of lakes, added a further argument: these birds were presumed to exist without access to food and air throughout their hibernation. In the second volume of his *Dissertation*, Jean-Jacques Bruhier gave a lengthy account of the strange habits of these birds, quoting observations from many older authorities.[12]

Another strange phenomenon was the "fasting girls" who claimed to be able to live for months, or even years, without taking any form of nourishment. In the sixteenth century, such cases had been presumed to be supernatural in origin. The decadelong fast of Margarethe Weiss of Speyer was the subject of a broadside in 1542. Noted for her piety, she claimed to take nourishment only from the transsubstantiation of holy wafers that floated down her throat. Barbara of Ulla, a Westphalian girl, said she had fasted for a year after being bewitched. Both of these girls were exposed as frauds after being put under strict supervision, but the reputation of the Dutch girl Eva Pfliegen, who claimed to have lived seventeen years sustained only by the scent of flowers, remains untarnished. In the late seventeenth and early eighteenth centuries, the fasting girls were no longer considered miraculous. Instead, just like the screaming cadavers, the toad in the hole, and the hibernating swallows, they became interesting and instructive phenomena of natural history and physiology. Several of them were described before various learned societies, foremost among them the young Englishwoman Martha Taylor, whose supposed yearlong fast was the subject of a discourse given before the Royal Society of London in 1669.[13]

Even more important for the growing anxiety about the uncertainty of the signs of death were the writings of Pechlin and Langelot, mentioned in the preceding chapter, concerning individuals who had stayed alive underwater for prolonged periods of time. The astounding observation of the submarine gardener Erik Björnsson can be verified from other, independent sources, and it lent an undeserved aura of credibility to some of Pechlin's other cases, like the tales of the Laplander who had once "drowned" for six weeks before being rescued and the merry young Gocellinus who had taken a fifteen-day dive before emerging alive and well, tales that belong wholly to the realm of fantasy. The Björnsson case even came to the attention of the Royal Society of London. The secretary of this famous body, Henry Oldenburg, wrote to the Swedish scholars Georg Stiernhielm and Urban Hjärne to obtain details of it, which the two Swedes willingly provided.[14] One of Hjärne's correspondents was Robert Hooke, and the case of the submarine gardener may well have inspired this noted physiologist to perform his famous experiment of "preserving animals alive by blowing air through their lungs with bellows," as he himself expressed it. Hooke was the first person to demonstrate that the supply of air to the lungs, and not the respiratory movements in themselves, was necessary for life.[15]

In a way, the 1740s were exactly the right time for fears of being buried alive to develop: there was still a good deal of superstition about abnormal fasts, hibernating swallows, and submarine humans, but the rationalist eighteenth-century medical scientists no longer considered these matters to be supernatural. Just as Bruhier had changed the interpretation of Kornmann's miracles of the dead into factual observations of people in extremis being buried alive, they used these observations to extend the limits of human physiology and to cast doubt on the prevailing signs of death, particularly the lack of respiration and arterial pulsations. Even more important, in 1740, the year Winslow's thesis was published, René-Antoine Ferchault de Réaumur published the earliest observation of artificial respiration, in a short work on how to resuscitate drowned people. That an unconscious, breathless, pulseless individual could be brought back to life by such means was nothing less than sensational, and it did more than anything else to increase the uncertainty about the currently used signs of death. The enterprising Bruhier was well aware of this and even included the entire text of Réaumur's paper in the 1742 volume of his *Dissertation*.[16] The preface of the English translation of Bruhier's book stressed artificial respiration as a major reason for doubting the reliability of the traditional signs of death.[17]

A no less important factor was also overlooked by earlier commentators: the personality of Jean-Jacques Bruhier. If Winslow was the typical scholar—quiet, unassuming, and unworldly—Bruhier was an entrepreneurial, flamboyant *homme du monde*. From 1742 until his death in 1756, he personified the anti-premature-burial agitation in France, and the collected edition of his books on the subject filled a small bookcase. Much of the increasing anxiety about the uncertainty of the signs of death, and the rapid upsurge in fear of premature burial in mid-eighteenth-century France, was due to his clever and ceaseless agitation. It would be unfair to accuse him of being guided by selfish motives—not even Louis did that—but the considerable fame he derived from his championship of burial reform cannot have been unwelcome for an ambitious, middle-aged man who had spent most of his previous life in relative obscurity.

JEAN-JACQUES BRUHIER was the winner of the debate on the uncertainty of the signs of death, and after the objections of Louis had been brushed aside, the majority of doctors and educated people agreed with his views on the subject. For example, the article on death in the famous *Encyclopédie* of Diderot and d'Alembert closely adhered to the teachings of Bruhier. The influential naturalist Georges-Louis Leclerc de Buffon was also influenced by Bruhier, and traces of his arguments can be found in the writings of the great physiologist Albrecht von Haller.[18] Bruhier's books reached a wide audience, and some of his readers wanted to carry on his campaign for burial reform; these volumes were to serve the anti-premature-burial activists as a repository of gruesome anecdotes for years to come. The first horror-mongering pamphlet, entitled *Terrible Torture and Cruel Despair of People Presumed Dead Who Are Buried Alive*, appeared in 1752, within Bruhier's lifetime.[19] By this time, the regulation of burial customs in rural France was left to the clerical authorities in the individual dioceses, and there is evidence that they were much aware of Bruhier's new theories, possibly as a result of his ceaseless efforts to promote his work at various academies. Quite a few French dioceses either made it obligatory to wait twelve or twenty-four hours between death and burial or started to enforce legalization that was already in place. As early as 1617, the diocese of Angers had stipulated that a period of twelve hours elapse before any person was buried, but it appears that Buhier's arguments changed the way this ancient law was interpreted. In 1755, the king's prosecutor in Angers took to court a woman called La Jumelière, whose work it had been to transport corpses to the cemetery, for burying people too soon. At the trial, one witness treated the jurors to a reading from a popular pamphlet on premature burial, about a man who had eaten his hands in despair when awakening in the tomb; another witness had himself seen a man stir in the local mortuary after having falsely been presumed dead. Considerable anger was directed at the wretched La Jumelière, whose ignorance and undue haste had exposed people to the danger of a horrible death underground.[20]

Particularly in France, at this time there also was a fear of the putrefactive fumes emanating from old churchyards. These mephitic exhalations were considered very detrimental to people's health, and from the 1740s onward they gained increasing prominence as a causative agent of disease.[21] The observations gathered to promote this view ranged from the factual—cemetery workers who fainted when entering old vaults far below ground, and contagion from corpses that had died in epidemics—to the ridiculous—that a putrid corpse poisoned an entire family of thirteen, or that all children who grew up near a cemetery became stunted and feebleminded. Medical as well as populist writing on this subject existed, along with appeals to move all cemeteries out

An anonymous eighteenth-century French drawing of a "corpse" awakening on the dissection table, frightening the anatomist. *Copyright the Wellcome Trust.*

from the cities to get rid of this serious threat to the public health. The appeals were answered by counterarguments: one was that several anatomists, who had spent all their working lives among putrid corpses, had became nonagenarians, one of them being our old friend Jacques-Bénigne Winslow. The fear of putrid exhalations also affected the appeals for burial reform. It was considered unrealistic to keep dead bodies above ground for several days awaiting putrefaction, since the gain in saving a few apparently dead people would be offset by epidemics of mephitic poisoning from the charnelhouses where the corpses were to be kept.

The French fear of being mistaken for dead was further fueled by a widespread tale about the death of the abbé Prévost, the famous author of *Manon Lescaut*. In November 1763, the abbé was found dead in the

Chantilly park; the cause of death was suspected to be apoplexy. He was taken to the house of a local surgeon, who proceeded to dissect the presumed corpse. But at the first incision, Prévost gave a loud cry and grasped the surgeon's hand. One version then says that he bled to death after the first stroke of the scalpel, another that he wrestled with the surgeon, broke free, and ran screaming from the house, but fell dead outside its gates. Both of these stories were disseminated by a writer named Antoine de la Place, who claimed to have received the details from the abbé de Blanchelande, Prévost's brother, but this gentleman denied the tale emphatically. The official *procès-verbal* of the abbé Prévost's death mentions nothing of any dramatic premature dissection. Posterity again preferred a good story to the truth, however, and the horrid death of Prévost was repeated by many anti-premature-burial campaigners.[22] The event even gave rise to an often repeated bon mot, albeit one unlikely to appeal to a medical-negligence lawyer. The philosophical M. de la Place, who claimed to have been a great friend of the late abbé Prévost, was asked by some of his other friends what they should do about the careless surgeon who had caused the abbé's unnecessary death. "Just sigh and remain silent," said M. de la Place in the spirit of true Christian forgiveness.

Several other countries took up the worries about premature burial. The Austrian court physician Gerard van Swieten agreed with Bruhier that putrefaction was the only certain sign of death. In 1756, two years after the German translation of Bruhier's book had been published, he proposed to the empress Maria Theresia that it become obligatory throughout Austria that at least forty-eight hours passed between death and burial. The corpses could be kept in communal morgues if the bereaved relatives did not want to have them in their houses.[23] Many German states followed suit and demanded a delay in burial of between twenty-four and forty-eight hours; Frankfurt and Saxony even extended this to three full days. This led to trouble with the Jewish population, whose time-honored practice had been to bury the dead as soon as possible. In 1772, a certain Dr. J. P. Brinckmann published a book arguing that premature burials occurred frequently in Germany; he quoted extensively from Jancke's translation of Bruhier and repeated many of the spurious old cases.[24] The same year, the duke of Mecklenburg-Schwerin issued a regulation that required Jews to allow at least three days to elapse between death and burial. There was considerable opposition to this proposal from the local Jewish groups. The influential Jewish philosopher Moses Mendelssohn wrote a letter to the duke suggesting that the regulation be changed to say that no Jew should be buried without death being certified by a physician, and the duke accepted this compromise. Mendelssohn had also written that, in his opinion, the ducal proposal violated no religious law, but the Orthodox Mecklenburg Jews disagreed. In 1787, the physician Marcus Herz, a close associate of Mendelssohn's, wrote a pamphlet arguing against early burials. He did not resort to religious arguments to convert his fellow Jews, preferring to frighten them with graphic descriptions of a man awakening in his coffin. By this time, the horror-mongering pamphlets of Brinckmann and other authors were on the market, and Herz had no lack of material with which to curdle his readers' blood. His pamphlet had the desired effect: growing minorities within the Jewish community formed burial societies that abandoned certain elements in their traditional ritual and that kept their dead above ground for two or three days.[25] It has also been demonstrated that in Belgium the time between death and burial was gradually extended to between one and two days during the 1770s and 1780s.[26]

Thus, one largely positive effect of Bruhier's work was that it became accepted practice, throughout large parts of Europe, to wait at least twelve or twenty-four hours before any person was buried. This was a compromise between the fear of premature burial and the similarly potent fear of mephitic poisoning from the corpses, thus producing a period of observation for individuals presumed dead, but not awaiting the full development of putrefaction. The demonstration of the uncertainty of the signs of death, particularly in the hands of careless, uneducated people, was heeded by medical practitioners all over Europe. Bruhier's words of warning that special care be taken with people thought dead from apoplexy, drowning, freezing to death, and "hysteric passions," were largely sound. Along with his directions for vigorous resuscitation of such presumed corpses, this advice probably saved many people from death, and some from being buried alive. The second positive effect of Bruhier's work was that his redefinition of death as a sometimes treatable, not always irreversible medical condition set the stage for the appearance of humane societies dedicated to saving people from apparent death. By the 1760s, it was clear to many physicians that a lingering death from chronic disease was a process very different from sudden death from drowning, asphyxia, seizures, or accidents; victims of these latter calamities could be resuscitated and brought back to life even if they had ceased to breathe. The first humane society was founded in 1767 in Amsterdam, where many people were lost in the canals every year. Venice, Milan, and Paris followed suit, and the Royal Humane Society was founded in London in 1774. Before long, Philadelphia, Massachusetts, and New York had humane societies as well. The societies were often founded and led by physicians, but they encouraged the general public to learn resuscitation techniques and gave medals and rewards to successful lifesavers.[27] In 1775, a Prussian society paid thirty marks for successful and fifteen for unsuccessful attempts to resuscitate people found in a lifeless condition. It paid no rewards when it suspected poisoning to be involved, probably to discourage abuse of the system by people who deliberately poisoned others, to gain money through resuscitating them. Doctors were also entitled to these rewards, but it was expressly stated that *prison doctors were not*; clearly, the philanthropic Prussians had no desire to perpetuate the criminal classes.[28] Although the humane societies and the anti-premature-burial propagandists were destined to go their separate ways in the 1800s, it is clear that a kind of cross-fertilization between these groups occurred in the late eighteenth century. The humane societies were obviously affected by the arguments for resuscitation of apparently dead people presented by Bruhier, Brinckmann, and others. At the same time, the public acceptance of the notion that a person who had ceased to breathe could still be successfully revived caused them to further doubt the definition of death, as explicitly discussed by some French pamphlet writers on premature burial.[29] The founder of the Royal Humane Society of London, Dr. William Hawes, was also the author of a tract on premature interment, in which he used the society's resuscitation case records to demonstrate the fallibility of the signs of death, except for putrefaction.[30]

But although the English establishment embraced the idea of humane societies, the vast majority of the medical profession turned a deaf ear to the eighteenth-century agitation regarding premature burials, and there was no question of extending the period between death and funeral because of these concerns. But already at this time, there is evidence that some English people left legacies to their family physicians to safeguard themselves against this gruesome fate. In 1769, Lord Chesterfield wrote, "All I desire, for my own burial, is not to be buried alive. . . ."[31] Miss Hannah Beswick, a wealthy spinster living in Manchester at about the same time, was even more frightened of being prematurely buried, after her brother had narrowly escaped this dire fate. She left 20,000 guineas to her family physician, Dr. Charles White, on the condition that she never be buried, that her body instead be embalmed and kept in the doctor's collection of anatomical preparations. Every day, for several years, the doctor and two reliable witnesses were to lift the veil and survey the countenance of this macabre resident patient for any signs of life. Later, the doctor put Miss Beswick's mummy inside an old clock case and opened the clock face once a year to see how his favorite patient was doing. Thanks to her eccentric will, the immortal Miss Beswick became quite a celebrity. As a child, Thomas De Quincey was taken to Dr. White's house in Lower King Street, Manchester, to see the mummy inside the old grandfather clock. The doctor loyally fulfilled his side of the deal to the end. After his death in 1813, Miss Beswick's mummy was taken to the Manchester Natural History Museum, where it was publicly shown in the Entrance Hall. Finally, it was decently interred in 1868.[32]

V

Hospitals for the Dead

Worn with old age and penury, nor thence
Rescued by any man's beneficence,
Into this tomb with tottering steps I past
And hardly here found leave to rest at last.
Usage for most doth after death provide
Interment, I was buried ere I died.
—Richard Garnett, "On One Who Died in a Tomb," from his Idylls and Epigrams *(1869)*

In 1787, the French doctor François Thiérry published a book in which he stated his conviction that most people did not die until some time after the onset of the traditional signs of death. To protect these still-sentient, still-living "corpses," he put forth a proposal resembling that of Bruhier in 1745. In all French cities, special waiting mortuaries should be built, and these institutions should receive all recently dead people and keep them under surveillance until the onset of putrefaction. Thiérry's book was translated into German in 1788.[1] That same year, the Austrian physician Johann Peter Frank published the fourth installment of his influential multivolume work on social and forensic medicine, in which the current debate about apparent death was thoroughly reviewed.[2] In his autobiography, Frank claimed that he had once himself been in a state of apparent death for four hours before recovering, and this naturally made him acutely aware of this problem.[3] Like the majority of European physicians of the time, Frank was a follower of Bruhier and convinced that putrefaction was the only reliable sign of death. He recommended that corpses be kept above ground for two or three days to await the onset of foulness. But Frank was against keeping the corpses in houses, since the overpowering stench would drive the living people from the household. In Toscana, a grand ducal edict of 1775 had ordered that all people be certified dead by a physician; the corpses were kept in dwelling houses, or even inside the churches, until the medical man arrived. Frank had obviously been there himself and experienced the nauseating stench that permeated these buildings. Without mentioning Thiérry's book, he suggested that a communal *Totenhaus*, a house for the dead, be constructed in every town, to allow the supervision of alleged corpses until they could safely be declared dead.

Johann Peter Frank was one of the most influential medical men in the German-speaking part of Europe, and the idea caught on. The first to act upon it was the twenty-nine-year-old Christoph Wilhelm Hufeland, a practicing physician in Weimar.[4] In 1790, he published an article in the magazine *Neue deutsche Merkur*, outlining his plan for a house for the dead to be erected in his hometown.[5] After the head of state, Duke Carl August, had given his gracious support to this novel idea, Hufeland gathered money by means of subscription among the wealthy. Countess Bernstorff, the widow of the former prime minister of Denmark, supported him.[6] A both practical and energetic man, and a true philanthropist, Hufeland personally supervised the planning and building of what he called the Vitae Dubiae Asylum, the asylum for doubtful life. In 1791, work on the mortuary was well under way, and Hufeland decided to write a pamphlet detailing his views both on real versus apparent death and on the practical aspects of saving people from premature burial by means of waiting mortuaries like the one in Weimar.[7] This was to become the most influential publication of its kind since the works of Bruhier.

Hufeland was a philanthropist, not a horror monger, and he did not accumulate hundreds of dubious and horrific case reports, as Bruhier and Brinckmann had done. He agreed with them that premature burial was a common occurrence and that putrefaction was the only certain sign of death. Indeed, he defined the death of an organism as the progressive disintegration of its structure: a slow, gradual process, complete only when total molecular decomposition had occurred. Time was the only competent judge between life and death, Hufeland wrote, and the power of life within the human frame was such that the only solution was to keep each corpse in a warm environment and to await the onset of putrefaction. He then proudly described the Weimar *Leichenhaus,* which looked rather like a normal dwelling house from the outside. Indoors, it had a corpse chamber with eight stretchers,

A lithographic portrait of Christoph Wilhelm Hufeland, by F. Krüger after M. Gavei. *From the author's collection.*

which could be watched by the attendant through a window. The keeper of the mortuary also had to stoke the fire in the kitchen, which heated water for a central heating canal underneath the corpse room. Hufeland sincerely hoped that cities all over Germany would follow the example that Weimar had set them: so low were the costs and so great the advantages of the new *Leichenhaus*. Hufeland had a low opinion of the *Totenweiber*, the female attendants of the old-fashioned mortuaries in various German towns: they were not only immoral but also incredibly stupid and ignorant. As a telling anecdote about their intellectual power, he recounted how one of them had once seen a "corpse" that tried to crawl out of his coffin. The *Totenweib*, fearful that he was the devil, had struck him with a broom and shouted, "Go back in there! What do you want with the living—you do not belong with us!"

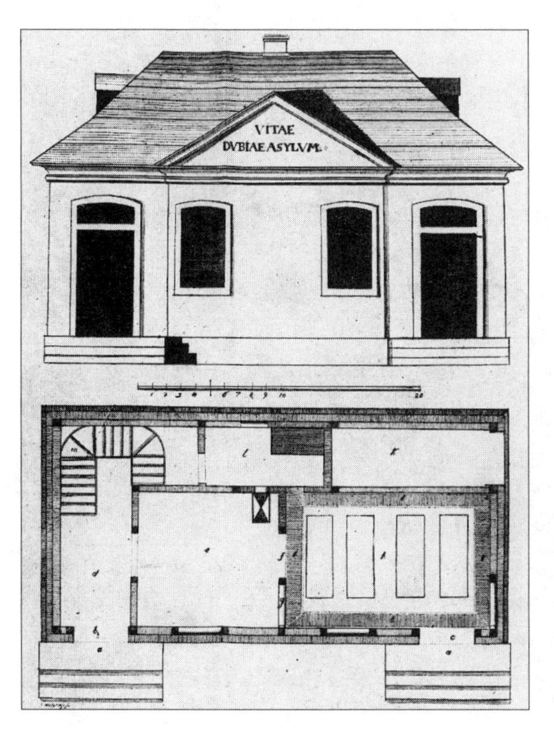

The plate from Hufeland's book showing the design for his Vitae Dubiae Asylum, the first waiting mortuary, which was built in 1792. *From the author's collection.*

Instead of these superstitious old crones, Hufeland wanted to establish a corps of young, well-educated men with all their senses intact, who should be employed as keepers of the *Leichenhäuser* that should be erected all over Germany. He had no suggestion how they should be enticed to take up such a loathsome profession, however. The low pay he had to offer was unlikely to tempt any well-trained young man to exchange a comfortable life in a wholesome rural atmosphere for the constant vigil by the window of the corpse room, engulfed in the putrid fumes of the *Leichenhaus*.

After Hufeland's pamphlet had been published, there were calls for waiting mortuaries to be constructed in many cities throughout the German states.[8] The outcome of these calls depended on whether the graphic descriptions of the horrors of premature interment in the proposals could appeal to the philanthropic zeal of the German burghers and overcome their natural reluctance to spend. It was not until Hufeland moved to Berlin in 1793, and personally canvassed the leading Prussian magnates with the assistance from a kindred spirit, the Dean Teller, that the foundations could be laid for the second *Leichenhaus*. This establishment, located in a cemetery just outside Berlin, was opened in 1795. Unlike that in Weimar, which relied entirely on the vigilance of the attendant, the Berlin mortuary had a system of strings tied to the fingers of the senseless inmates; these strings were connected to a large bell.[9] Another, more luxurious waiting mortuary was built in Berlin in 1797, for the considerable sum of 3,000 thaler: it had one chamber for male and one for female corpses. In the years that followed, further *Leichenhäuser* were built in Brunswick, Ansbach, Kassel, and Mainz. The most luxurious to date was erected in Munich in 1808. It used not sex but financial wealth as the criterion of division between its two corpse chambers: it cost just two gulden for a corpse to be admitted to the common section, but five times as much to enter the luxury section of the mortuary. Some philanthropists thought this was a betrayal of Hufeland's noble idea: Did these wealthy snobs imagine that the Kingdom of Heaven would also have separate compartments for the wealthy and the poor? The Munich *Leichenhaus* also came under fire for admitting visitors: any person could, for a small fee, come in for a stroll among the lavish mortuary's many pictures, flower displays, and statues, although the putrefying corpses were, of course, the main attraction. The system to detect revival of the inmates in the Munich *Leichenhaus* was both novel and controversial: the strings from the fingers and toes of the corpses were connected to a large harmonium with air-pressured bellows. Every day, the mortuary attendant played this harmonium to demonstrate that it was fully functional. At night, the swelling of the putrefying corpses frequently set off the easily triggered mechanism, however, and the attendant was awakened by a ghostly symphony emanating from the corpse chamber.[10]

IN THE 1790S, as concerns about apparently dead people grew widespread in the German states, a new literary genre was born. As a strange side effect of the philanthropic concerns, several "amusing" collections of anecdotes and short stories about apparently dead people either fortuitously awakening in their coffins or suffering superlative torments in their premature tombs were published, and had no difficulty in finding readers. These books were devoid of philanthropic ambitions and did not argue for or against waiting mortuaries; their only aim was to make their reader's flesh creep through a mixture of burlesque amusement and gloating sadism. They claimed to contain true stories, but gave no references to their sources, and much of their content was no doubt pure fiction. In one of these works, the anonymous *Wiederauflebungs-Geschichten scheintodter Menschen,* we again make the acquaintance of the Lecherous Monk, the Careless Anatomist (twice), and the Two Young Lovers (twice). The Lady with the Ring appears in three incarnations, from Paris, Leipzig, and London. Another burlesque anecdote concerns two students employed in a convent to maintain a vigil all night over the body of a recently deceased monk and to say prayers for his soul. The other monks give them wine and beer to keep them in good cheer, but the students drink too deeply and leave their prayer books unread and their strings of beads untouched. As a joke, one of them carries the dead monk onto his chair and himself lies down on the corpse's mattress. When the other student comes back, he sees that the presumed corpse moved one foot, and he runs to tell his friend. "But—Jesus Maria!—this was not his friend, but the Monk—and still worse, this Monk stared at him with his eyes wide open, and tried to get up!" The student faints, believing the living corpse to be a ghost. When the prankster sees the monk coming toward him, huffing and puffing from the exertion of moving his huge body, he falls dead from the shock. The monk, whose two corpse watchers lie outstretched on the floor in front of him, manages to call his colleagues; it turns out that one of the students can be revived, but that the wicked prankster is stone dead. Death must have his victim, and there must be a moral even to a vulgar, silly story like this one.

Had the *Wiederauflebungs-Geschichten scheintodter Menschen* only contained this kind of buffoonery, it would have done less mischief; but interspersed among the reviving monks and swooning ladies were the most detailed and gruesome horror stories of prematurely buried people—particularly women—awakening in the tomb. In two stories, prematurely buried women eat their fingers in their agony; in two others, they give birth to children in their coffins. A tale from Strasbourg combined these horrors. A young woman dies during pregnancy and is buried in a vault. Some time later, the vault is opened to admit another corpse. "But what a sight! The unfortunate woman had crawled out of her coffin and given birth to a child. Her body was lying on the ground, with the child in her arms, and its little fingers in her mouth, as if she had wanted to eat them. Here, after a superhuman struggle, the tyrannical torments of hunger had been vanquished by the tender motherly love." In a tale from Portugal, the vault of an ancient family is opened by workmen, and it is seen that one of the coffins is empty. In the basement of the vault, the workmen find an emaciated eighty-year-old man, with long, crooked nails and a beard reaching to his knees. This creature seems incapable of speech, and cannot stand the light from the torches. After being carried away and put to bed, he tells a remarkable story. Many years earlier, he had been deemed dead after a violent malady and buried alive in the vault. He kept alive by licking an acrid fluid from the vault wall, and by drinking the blood of insects and eating some noxious fungi that grew from the vault walls. "With these nails, I buried my excrements. . . . A Frog was my companion. He rested at my bosom, and followed me everywhere." If the reader had not had enough horrors from this idle tale, the next one, also from the Iberian Peninsula, was likely to do the trick. When an ancient burial vault in Spain is opened, a skeleton is found on the floor. It appears that before succumbing to his torments, the wretched creature made a lengthy inscription onto one of the tin coffins in the vault. He had sunk into a state of *Scheintod*, and although he heard everything that was going on, he could not prevent being buried alive. He wakes up in the vault, with no chance of succour. His desperation is closely and sadistically detailed: "The lungs almost fail to breathe the pestiferous air; my steps falter and my knees shake; my saliva becomes acrid and burns like phosphorus. I drink my own urine and eat my own excrements. On all four, I crawl to this coffin to write my story, under the most hideous torments. . . . Ah, what convulsions! What burning heat in my entire being! A foul-smelling fluid drips down from the window, and I wish to drag my frame over there to lick it from the marble, but my feet fail to obey me. Ach, if I could only open these coffins! Maybe some flesh still remains on the corpses they contain, flesh that would prolong my life. . . . Ah, what pain! Holy Saviour, see me in my misery! God, to whom nothing is impossible, save me! Save me! Save me!"[11]

A certain Heinrich Friedrich Köppen, a native of Halle in Germany, incorporated large parts of the *Wiederauflebungs-Geschichten scheintodter Menschen* into a book of his own in the same vein, published in 1800.[12] Köppen separated the stories into three categories: credible, less credible, and wholly incredible, and thinly veiled the gloating sadism of some of the stories with a philanthropic appeal to build waiting mortu-

Hier ruft, o möchte Gott es geben!
Wol schon ein Freund mir dankend zu:
Heyl sey Dir! denn Du hast mein Leben,
Vom Tode mir gerettet Du!
O Gott, wie muß dieß Glück erfreun,
Der Retter eines Freundes seyn!

Achtung
den
Scheintodten.
zum
Besten der Menschheit
herausgegeben
von
Heinrich Friedrich Köppen.

Erster Theil. Erläuterungen.

Noch viel Verdienst ist übrig. Auf, hab' es nur, die Welt wird's kennen.
Klopstock.

Halle,
bey dem Herausgeber.
1800.

The title page and frontispiece of Dr. Köppen's collection of amusing and gruesome anecdotes about *Scheintod*, with a stirring poem praising the philanthropic zeal of the *Leichenhaus* keeper. *From the author's collection.*

aries. The contents of another, similarly pleasant volume, the *Wirkliche und wahre mit Urkunden erläuterte Geschichten und Begebenheiten von lebendig begrabene Personen*, appear to have been gathered from the same source as those communicated by Köppen and in the *Wiederauflebungs-Geschichten scheintodter Menschen*. These collections of idle stories were avidly read throughout Germany and were sometimes culled for source material by the serious writers on the subject.

IN 1808, CHRISTOPH Wilhelm Hufeland summarized the situation seventeen years after his original appeal in a new book, entitled *Der Scheintod*, which was intended to be a complete encyclopedia for those interested in this subject. Hufeland suggested that a particular state of deep unconsciousness, the *Scheintod*, or death trance, always precluded real death. The *Scheintod* was indistinguishable from real death and could last for days or even weeks, like the hibernation of an animal. Although the apparently dead person seemed to lack arterial pulsations, muscular reflexes, and respiratory movements, the death trance was not always fatal; through vigorous resuscitation, the individual's life could still be saved. The onset of putrefaction was the only way of determining the death trance from real death. It seems that strange, trance-like states of this kind really existed at the time and that they were more common than today. The victims were usually younger women. Their perception was typically unimpaired, enabling them to hear and understand what was happening, although they were powerless to move or cry out, even if they were put in a coffin and prepared for burial. Sometimes, quite trivial matters aroused them from the trance. According to one of Bruhier's stories, repeated by Hufeland, one apparently dead lady woke up from her trance after hearing the voice of a childhood friend; another, after a lutist played one of her favorite melodies. M. Chevalier, a Paris surgeon, known as a great piquet player, was aroused from a deep trance, diagnosed by some as death, after one of his friends called out the piquet commands "Quint, quarante, point!" Hufeland received an even more remarkable case from his colleague Dr. Cammerer. This practitioner was one of the doctors attending the wife of a professor in Tübingen, who had collapsed after being seized by convulsions in the sixth month of her pregnancy. The woman did not move and could not be seen to breathe, and no pulsations could be felt. Application of a powerful itching powder produced no movement. The other doctors presumed that she was dead, but Dr. Cammerer applied a strong mustard plaster to the soles of her feet and tore it off with such force that parts of the skin came off as well. The woman's mouth twitched with pain. This prompted the doctors to attempt further brutal stratagems of resuscitation: they administered acrid enemas, thrust sharp needles under the nails, and applied glowing irons to the most sensitive parts. The professor's wife remained motionless and still. The absence of putrefaction necessitated further observation, however, and on the sixth day of her coma the woman opened her eyes. She then gave birth to a dead child and afterward recovered completely. Hufeland did not mention whether or not she was grateful for the medical attention she had received during her illness, tortures that would not have seemed out of place in the dungeons of the Spanish Inquisition.

In his new book, Hufeland wrote that the philanthropic citizen should feel a sense of elation, not revulsion, when handling a presumed corpse, since the "deceased" might still hear and feel what was going on and secretly bless his or her savior. He heaped praise on Jean-Jacques Bruhier for his pioneer work on behalf of the apparently dead. On the other hand, he blasted Antoine Louis's book for being full of errors, saying it did not even come close to refuting Bruhier's arguments. Hufeland was pleased to note that since the publication of his own pamphlet in 1791, no fewer than twenty-six German books and pamphlets had been added to the bibliography of apparent death and premature burial. Indeed, this problem was given more attention than almost any other medical topic of the time. There was more sympathy for the apparently dead than for the starving, the sick, the crippled, and the destitute. All over Germany, mealy-mouthed and sanctimonious clergymen, inventors, and philanthropists grasped their pens to defend the helpless victims of apparent death. They drew up grandiose schemes for huge waiting mortuaries, manned by armies of watchmen, and invented the most strange and impractical security coffins for use as a safeguard against premature burial. The vast majority of these writers agreed with Hufeland that putrefaction was the only certain sign of death and that waiting mortuaries were needed; in glowing terms, they praised his philanthropic efforts for the benefit of the apparently dead. At this time, there flourished a culture of sensibility and an awareness of those less fortunate in life. In England, prostitutes were put in homes for "unfortunates," beggars taken off the streets, and humane societies supported; in Germany, the philanthropists concentrated on the greatest misfortune of all—that of being buried alive. When a certain Pastor Wolff delivered an eloquent sermon on apparent death and waiting mortuaries in Brunswick in 1792, there were ecstatic scenes as he was congratulated by people rushing forward to donate money for a *Leichenhaus*.[15] It is apparent that an almost religious fervor came over the Germans who devoted their efforts to saving the helpless victims of the death trance: the predicament of the still-sentient, but senseless and paralyzed, person, who could not prevent that he or she was shrouded, put in a coffin, and buried, seems to have affected many educated people very strongly.[16] As hinted by the pamphlets quoted earlier, this perverted zeal for philanthropy was not unaffected by darker undercurrents of necrophilia and sadism.

The first sign of dissent from Hufeland's views was a pamphlet by a Dr. Johann Daniel Metzger, published in 1792.[17] Metzger did not deny that apparent death was a serious medical problem, but objected that the deceased should be kept in their own homes rather than among the mephitic fumes of the *Leichenhaus*. He himself would almost rather be buried alive than awaken among the putrid corpses in what he called Hufeland's house of horrors. In his new book, the philanthropic Hufeland magnanimously reviewed Metzger's erroneous opinions, adding that few later writers had agreed with him. Hufeland was more concerned with another, more ribald publication, issued in Prague by a certain von Müller.[18] Entitled "How Prematurely Buried People Can Easily and Conveniently Be Helped out of Their Graves," this pamphlet attempted to make fun of the contemporary obsession with apparent death. It is a breath of fresh air compared with the repetitive, sanctimonious books on the subject and provides much-needed evidence that at least one German writer on the subject had a sense of humor. In his tongue-in-cheek pamphlet, Herr von Müller suggested that all corpses be put in coffins with lids made of glass and that a strong hammer be placed in their hands. These coffins should not be buried, but kept in large vaults under the churchyard. Upon awakening from his death trance, the individual could easily smash his way out of the coffin, and dispose of the glass door to the vault in a similar manner, before emerging to greet the watchmen patrolling this model cemetery. Hufeland wrote that he had wept, not laughed, when reading this "funny" pamphlet; he ended his review by asking the forgiveness of his readers for acquainting them with this sorry attempt to deride and debase the noble and philanthropic *Leichenhaus* movement.

Hufeland intended *Der Scheintod* for a wider readership and probably feared that the lengthy sections on reviving drowned people, and the reviews of other publications on apparent death, would scare lay readers away. He decided to spice it up with various horror stories, some fetched from the *Wiederauflebungs-Geschichten scheintodter Menschen* and others from an even more dubious source—a collection of ghost stories called Waginger's *Neue Gespenster*.[19]

HUFELAND'S ORIGINAL *LEICHENHAUS* in Weimar had been strictly utilitarian: his "Asylum for Doubtful Life" was like a dwelling house for the dead, with the necessary equipment to supervise and resuscitate them. But already in 1796 the architect Jacob Atzel had proposed that the waiting mortuaries be regarded as important sacral monuments, just like a church or a mausoleum.[20] The mortuaries of Ansbach and Munich were constructed along these lines, but even more ornate. In particular, the German *Leichenhaus* architects favored the style of the Italian architect Andrea Palladio and his Villa Rotonda; some of their hospitals for the dead were built in marble and had grand columns and ornate façades and cupolas. In the landscaped gardens of the necropolis, surrounded by sphinxes and statues, the waiting mortuary symbolized the transfer from life to death; the exit from this Temple of Sleep led to the world of the dead.[21]

The most luxurious of the newer waiting mortuaries was opened in Frankfurt am Main in 1828. Built from a design by the architect Johann Michael Voit, it had two wards of twenty-three beds each, one for male and one for female corpses; the lifeless "patients" had their hands and feet connected to an intricate system of strings, leading to a clearly sounding bell. These wards were well aired and, during winter, centrally heated. Several watchmen were employed there, and a resident doctor did his rounds every day. There was a kitchen for the watchmen, a boiler room, a bathroom and a resuscitation room well equipped with surgical instruments, and a complete pharmacy.[22] The cemetery director had written a *Todtenwärter-Ordnung*, an instruction manual for the watchmen who lived in the mortuary, which reveals that although these individuals benefited from light duties in an indoor climate, the discipline was strict and the duties wholly unprepossessing. The watchmen were not allowed to smoke, curse, or drink alcoholic beverages in the mortuary, nor could they receive visitors. They worked in shifts and had to obtain written permission from the director if they wanted to go out of doors, even for twenty minutes. Their communal bedroom had a large clock outside it; the watchman had to set it before going to bed, so that the director could see how long he had slept, and punish those who tried to have an unauthorized lie-in. Apart from supervising the corpses, the watchmen were perpetually employed in cleaning the corpse beds and various other utensils used, in a vain attempt to combat the stench spread by the putrefying bodies.

In 1834, the Weimar practitioner Carl Schwalbe published a book that neatly summed up the first forty-five years of German waiting mortuaries.[23] It was dedicated to the great founder of the movement, Christoph Wilhelm Hufeland, still alive at the age of seventy-two. His old *Leichenhaus* in Weimar had been destroyed by marauding French troops during the Napoleonic Wars, but a new one had been built in 1824 at the considerable cost of 2,500 thaler. This new Weimar waiting mortuary was considered a model of its kind, and Schwalbe took care to explain its workings thoroughly, hoping that this would lead to the erection of similar buildings abroad. His book has three excellent plates of the equipment. The corpse beds, strings, and bells were similar to those used in Frankfurt, although Schwalbe was otherwise unimpressed by the calls from aesthetes and architects that the waiting mortuaries be considered parts of sacral architecture. The German *Leichenhäuser* were

Hufeland's legacy to the people, he wrote, and his original plan for their construction should be adhered to as far as possible. Dr. Schwalbe treaded lightly when describing the actual decomposition of the corpses in the warm, fetid room of the *Leichenhaus*, except to point out that early burial was necessary for poor people whose corpses were swarming with all kinds of vermin. Dr. L. Aug. Kraus supported this notion, pointing out that in cases of the dreaded lousy disease, in which the sufferer was literally eaten alive by insects living in tumors under the skin, very early burial was necessary in order to prevent contagion.[24] Dr. Kraus also suggested that every person be autopsied at the *Leichenhaus* before burial, both for the sake of medical science and to make sure that no one awoke in the grave. In 1834, an elaborate warning system, with strings connecting the fingers and toes of every inmate to an advanced clockwork made by the watchmaker Herr Zachariae, was constructed for the Leipzig *Leichenhaus* at the order of the chief physician, Dr. Clarus.[25]

But already during Hufeland's lifetime, there were ominous signs that all was not well with the *Leichenhaus* movement. Although doctors like Schwalbe and Kraus, and architects like Atzel and Voit, were still enthusiastic, the German clergy, who as we saw had been instrumental in supporting Hufeland in the 1790s, gradually lost faith in the waiting mortuaries. This loss is likely to have been due to a combination of two independent factors. First, it must have been evident to the most enthusiastic philanthropist that these waiting mortuaries were smelly, unpleasant buildings, and more than one parson must have considered it an abomination that polluted his churchyard. Second, there were calls from medical writers that a grand scheme of waiting mortuaries in every German town be *financed by the church,* and this is likely to have dampened the enthusiasm of the senior clerics concerned. Among the ordinary people, there was also a widespread feeling that the *Leichenhäuser* had outstayed their welcome. Unaffected by the horror stories of premature burial that still circulated in newspapers and popular pamphlets, they resolutely refused to give up their dead and bring them to these foul-smelling hospitals for the dead. In

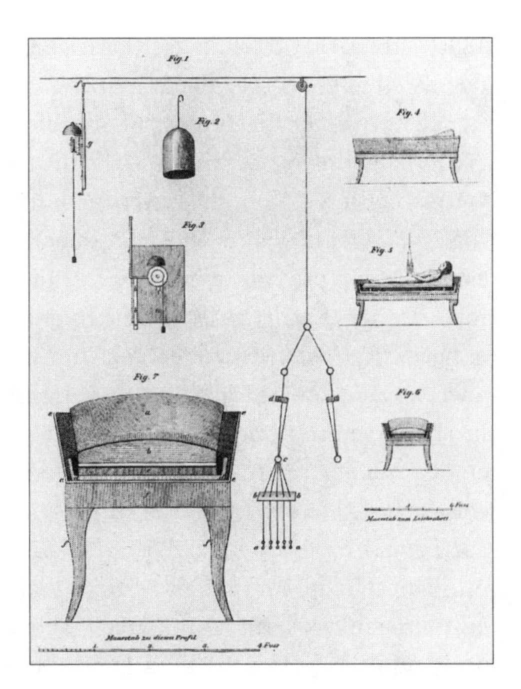

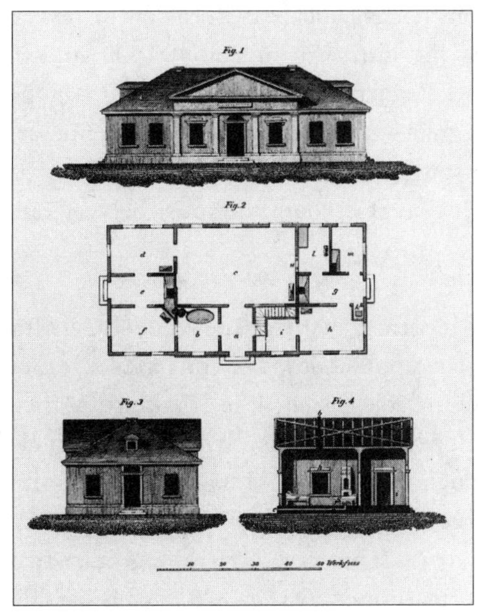

The exterior of the second Weimar waiting mortuary, and a design for its corpse beds, strings, and bells; two plates from Dr. Carl Schwalbe's *Das Leichenhaus in Weimar* (Leipzig, 1834).

1830, Professor Carl Ludvig Klose, of Breslau, wrote that during the twenty-nine years that the *Leichenhaus* in that town had been operational, only nineteen corpses, all of well-to-do citizens, had been admitted for supervision within its walls. In Freiburg, and probably in other German cities as well, there were rumors among the lower classes that the *Leichenhaus* was a bluff and that the corpses brought there were clandestinely used for medical experimentation. In 1842, only one of the six waiting mortuaries in Berlin remained in use, because of a lack of "patients." A visiting French doctor wrote that less than one corpse a week was admitted to this sole *Leichenhaus* in the Prussian capital. In 1837, even the grand waiting mortuary in Frankfurt am Main stood empty much of the time; not one in ten of the deceased ended up on its corpse beds. Many of those corpses saved from exhibition in the *Leichenhaus* were being deliberately withheld by "lower-class people, into whose stupid minds the idea of incubating corpses until they are putrid has never been assimilated," as a certain Dr. Graff contemptuously put it.[26] A Dr. von Steudel wrote that in the years 1828–1849 the mortuaries and other precautions to save apparently dead people in the state of Württemberg had cost the immense sum of 400,000 gulden. Approximately a million corpses had passed through the system; none of them had awakened in the mortuaries. Although Dr. von Steudel would not commit himself to stating that there was no such thing as the *Scheintod*, he had to admit that this condition must be very rare indeed.[27]

BY THE MID-NINETEENTH century, many of the smaller German waiting mortuaries had either been pulled down or converted to other uses. On the other hand, those in larger cities, like Frankfurt, Mainz, Munich, and Berlin, were still operational. Actually, quite a few new *Leichenhäuser* were built throughout the second half of the century: the one in Mainz was reconstructed in 1865, the one in Ulm in 1872, and the one in Düsseldorf in 1875. The city of Hamburg built two *Leichenhäuser* in 1871, and Stuttgart built two in 1875 at considerable expense. In 1871, the city of Lemberg decided to pull down its old *Leichenhaus*, which dated back to 1802, and erect a new one with an electric warning system for the corpse beds in its two corpse rooms.[28] Frankfurt had two waiting mortuaries in the 1880s: the one built in 1828 and one that had been rebuilt in 1848 and later enlarged. This latter establishment boasted a monumental façade and was particularly impressive. The design was a novel one: it did not have a large corpse room but individual funeral cells, each with a corpse bed and strings leading to the powerful alarm bell. This allowed for much needed improvements in ventilation, as well as privacy for the dead. Clusters of eight funeral cells were grouped around each of the watch rooms, with large windows to each cell surrounding the watchman. The advantages in ventilation and privacy were thus achieved at the price of having to employ quite an army of watchmen. This did not mean that the working conditions of these wretched watchmen had improved in any way, however. In order to make unauthorized naps more difficult, the watch room had no bed, table, or chair, and a crank handle had to be turned every half-hour to keep a powerful alarm from going off that would announce this neglect of duty. A visiting Frenchman considered the construction of the Frankfurt waiting mortuary to be by far the best he had seen during his tour of the German *Leichenhäuser*: he contrasted this clean, well-manned hospital with the charnelhouse atmosphere of the Mainz waiting mortuary, where overripe, bloated corpses were lying about in the large corpse room.[29]

In spite of the superior construction of the Frankfurt *Leichenhaus*, it was difficult to persuade people to deposit their dead there. In 1865, a French writer was appalled to learn that only one hundred corpses were admitted each year; the large, well-staffed mortuary was built for a far greater number.[30] There was a fearful cholera epidemic in 1869, however, and the civic authorities actually made it obligatory for every corpse to be taken to the *Leichenhaus* before burial. This system was kept up after the cholera epidemic had abated, in spite of objections from the burghers, who did not want the corpses of their dead relatives

A funeral cell from the Frankfurt am Main *Leichenhaus*. *From Dr. Josat's* De la mort et de ses caractères *(Paris, 1854).*

to be taken away from them, and the poor, who dreaded that anatomists would take liberties with the corpses. By the 1880s, it had become generally accepted that every dead body should be supervised for signs of life at the waiting mortuary before it could be buried. This was also the case in Munich, where several *Leichenhäuser*, one of them reserved for those of the Jewish faith, were operational as late as the 1880s and 1890s. A French author tells us that in the 1880s Munich had six waiting mortuaries, which were obligatory but not particularly popular among the citizens.[31] The Northern Mortuary was particularly impressive, with its three rows of twenty ornate wooden corpse beds, on which the inmates rested on slabs supplied with a zinc trench filled with an antiseptic solution. The head was raised and put on a cushion; the hands were crossed upon the breast and one finger inserted into a ring tied to a string communicating with the central alarm bells. There was a room for the rich and a room for the poor; the former had the more elegant and profuse floral tributes. During the either forty-eight or seventy-two hours each body was kept in this hospital for the dead, they were supervised by the mortuary attendant, whose humble, narrow cell had windows on all sides, so that he could not for a moment escape the sight (and smell) of his silent patients. Every hour, this wretched individual had to make a round to inspect all the corpses; he was not allowed to leave the mortuary even for a moment; and during the night, he was often awakened by a strident peal from the powerful alarm bell, since the least movement of the putrefying corpses set the mechanism in motion.[32]

Several accounts state that the waiting mortuaries of Munich and Frankfurt were actually open to visitors. Either the civic authorities found the entrance fees a welcome contribution to the considerable

The exterior of the Munich *Leichenhaus* in the 1890s. *From W. Tebb and E. P. Vollum,* Premature Burial and How It May Be Prevented *(London, 1905).*

The main room of the Munich *Leichenhaus*, showing the empty corpse beds with their flower decorations, and the arrangement of the strings leading to the central bell. *From W. Tebb and E. P. Vollum,* Premature Burial and How It May Be Prevented *(London, 1905).*

Another view of the main room of the Munich *Leichenhaus*, with several corpses connected to the rescue system with the strings. The strong-smelling flowers have the dual role of masking the stench of the putrefying corpses and of adding a decorative element to what would otherwise have been a gruesome scene. *From W. Tebb and E. P. Vollum,* Premature Burial and How It May Be Prevented *(London, 1905).*

costs of upkeep for these vast mortuaries, or they hoped that foreign tourists would be so impressed with these effective, modern establishments that they would plant the seed of Hufeland's great work in foreign ground. The mortuaries were part of the sight-seeing spots of these towns, and the foreigners who saw them did not go away unimpressed, although the prevailing sentiment was that of horror and disgust rather than of admiration. On a visit to Frankfurt, the British author Wilkie Collins was so impressed by the ghoulish hospital for the dead that he later made it the location for one of his horror novels.[33] An American journalist who had visited the Munich mortuary wrote that he would never be able to forget the sight of the long rows of bodies dressed in their wedding finery, with the wires attached to their hands and feet, or the anxious faces of the watchmen looking in through the windows. He ended his article with these words: "I pity the stranger who dies within the gates of Munich. Every one here is treated with equal injustice, be he high or low, friend or foe."[34] In the early 1880s, Samuel Langhorne Clemens, better known as Mark Twain, visited one of the two mortuaries still operating in Munich. It impressed him strongly, and his description should be quoted verbatim:

> It was a grisly place, this spacious room. There were thirty-six corpses of adults in sight, stretched on their backs on slightly slanted boards, in three long rows—all of them with wax-white, rigid faces, and all of them wrapped in white shrouds. Along the sides of the room were deep alcoves, like bay windows, and in each of these lay several marble-visaged babes, utterly hidden and buried under banks of fresh flowers, all but their faces and crossed hands. Around a finger of each of these fifty still forms, both great and small, was a ring, and from the ring a wire led to the ceiling, and thence to a bell in a watch-room yonder, where, day and night, a watchman always sits alert and ready to spring to the aid of any of that pallid company who, waked out of death, shall make a movement—for any, even the slightest, movement will twitch the wire and ring that fearful bell! I imagined myself a death-sentinel drows-

ing there alone, far in the dragging watches of some wailing, gusty night, and having in a twinkling all my body stricken to quivering jelly by the sudden clamor of that awful summons![35]

It was not until 1898 that the Munich waiting mortuaries removed their life-detection apparatus, by order from the civic magistrates.[36] Some other German mortuaries kept the alarm systems even longer. In the 1940s, the mortuaries of a large hospital in Alsace and the main mortuary in Strasbourg both had electrical life-detection systems, a push switch connected with the main alarm bell being put into the hand of each of the corpses as they were received.[37]

IN SPITE OF the calls from Hufeland and Schwalbe, few waiting mortuaries were built outside Germany. In France, the debate over whether or not to construct waiting mortuaries was to last more than a hundred years. François Thiérry's book advocating their construction had found many interested readers. One of them was Mme Suzanne Necker, the wife of Louis XVI's minister of finance Jacques Necker. She had once visited a hospital in Paris where the porters were either callous or careless enough to put dying patients into coffins instead of their hospital beds. Deeply affected by this reprehensible practice, Suzanne Necker imagined what could happen if the poor wretch in the coffin was not really dead when buried. She spent the rest of her life in fear of being buried alive. In 1790, she followed Thiérry in writing an appeal to build waiting mortuaries, but her pamphlet was without success.[38]

Count Leopold von Berchtold, having already called upon the emperor of Austria-Hungary and several other European sovereigns, put a proposal to build waiting mortuaries in France before the National Assembly in 1792.[39] In a country ravaged by revolutionary unrest, this proposal achieved little more than that of Mme Necker. But the concerns for the apparently dead continued during the revolutionary regime. In 1794, a certain Citizen Avril appealed to the General Council of the Seine that nine waiting mortuaries be built in Paris. In 1799, Citizen Cambry presented a more elaborate proposal. Four *temples funéraires*, elaborate waiting mortuaries that were ornately and elegantly designed, but devoid of every religious symbol, should receive all the dead in Paris. The temples should be well equipped to supervise and resuscitate apparently dead people.[40] This idea was approved by the revolutionary government, which liked its antireligious overtones, and the prefect of Paris made plans to build six of these temples in the French capital. But this was just before Napoleon Bonaparte became first consul. Although both M. Fiochot, the prefect of Paris, and Dr. Dessartz, the doyen of the Faculty of Medicine, prevailed upon him to build the mortuaries, Napoleon scrapped the idea, and no further proposals to build waiting mortuaries took place during his rule. In the Code Napoléon, it was instead stipulated that there be a twenty-four-hour delay between death and burial throughout the French Empire.

In Denmark, there was a good deal of agitation for the construction of a waiting mortuary after Hufeland's original appeal. In 1793, using the argument that what was necessary in Weimar and Brunswick was also necessary in Scandinavia, the clergyman D. C. Bastholm called for a waiting mortuary in Copenhagen.[41] It should be built as a model of Hufeland's original building, and to save money, the two watchmen should work twelve-hour shifts around the clock. Parson Bastholm hoped that a public subscription would pay for the erection of this first Danish *Leichenhaus*; he rather ominously added that the person who was guilty of the sin of betting more money at cards than it took to subscribe to the mortuary could hope for no sympathy if he or she awoke in a narrow coffin underground. But in spite of this sinister clergyman's appeal, it was not until 1808 that a waiting mortuary was built, at the Assistens cemetery of Copenhagen. It was equipped with corpse beds, strings and bells, and an attendant who blew a trumpet next to the unprotected ears of his charges to make sure they were not shamming death.[42] This mortuary was used until about 1850, but the life-detection system was scrapped at a much earlier date. Proposals to build further waiting mortuaries in Denmark and Sweden were not acted upon.[43]

In Austria-Hungary, Hufeland's appeal had been widely noted. The first to act upon it was the physician Adalbert Zarda, who had a small waiting mortuary built in Prague in 1797.[44] There is evidence that both Vienna and some smaller cities had waiting mortuaries in the 1830s. In 1860, after an appeal by the pediatrician Franz Hügel, a new waiting mortuary was built at the Zentralfriedhof in Vienna; it was still operational in 1874, and probably even longer.[45] It had one large corpse room full of corpse beds, rather like a hospital ward. The Vienna *Leichenhaus* was ultramodern even compared with the leading German mortuaries, and its corpse beds were equipped with electric contacts and bells of

Illustrirtes Wiener Extrablatt.

№ 273. Wien, Montag, 5. Oktober 1874. 3. Jahrgang.

Die Todtenhalle des Zentralfriedhofes.

Die heutige Nummer mit Inbegriff der Beilage ist 10 Seiten stark.

The corpse hall of the waiting mortuary connected with the Vienna Zentralfriedhof in 1874, from the *Illustriertes Wiener Extrablatt* of October 5, 1874. *Reproduced by permission of the Österreichische Nationalbibliothek, Vienna.*

recent manufacture. These were presumed to be less prone to false alarms, and the mortuary attendant could sit watching a large frame with little electric bells under indicators for each of the corpse beds, rather like a hotel porter waiting for one of the guests to ring for room service. There was a smaller corpse chamber for suicides, which conspicuously lacked the electrical warning system.

In Italy, there were several appeals in various cities to build waiting mortuaries, but nothing came of them. Nor was any mortuary of this kind ever built in Spain. According to one account, a waiting mortuary existed in Lisbon in 1812, but was stated to have been a short-lived venture prompted by some epidemic of contagious disease rather than by fear of premature burial. The 1840s saw a great upsurge in fear of premature burial in Lisbon. One account noted that riots in the streets erupted when demands for waiting mortuaries were not met, and the minister Costa-Cabral had to resign as a result of this uproar.[46] A Dutch pamphlet, published in 1837, claims that in addition to numerous German cities, Paris, Lisbon, and Copenhagen had waiting mortuaries at this time.[47]

The appeals of Hufeland and Schwalbe also reached Belgium and the Netherlands. Brussels is said to have had a waiting mortuary with fourteen corpse beds, which was constructed by Dr. Janssens, director of the Bureau of Hygiene, and situated near the St. Catherine church. It was operational as late as the 1870s, which time there were also two waiting mortuaries in Amsterdam.[48] The Hague had three waiting mortuaries at various times, of which the *Schijndodenhuis* (House for the Apparently Dead) in the Algemeen Begraafplaats was the grandest. Constructed in 1829–31, this elegant building was obviously inspired by the German notions about sepulchral architecture in the style of Palladio. It had a life-detection apparatus with bells and strings, which was used for only a very limited period of time, however. The *Schijndodenhuis* served as an ordinary mortuary for many years, and was recently restored and altered to become an office building. Its exterior has not been greatly changed since the 1830s, and it may well be the unique survivor of its particular kind of sepulchral architecture.[49] Some of the old German *Leichenhäuser* were still standing well into the twentieth century, although converted to chapels or ordinary mortuaries: the one in Worms was destroyed in 1945, and the one in Speyer was intact as late as 1959.[50]

There is a very curious note in a German journal about a waiting mortuary in New York that was supposed to be active in the 1820s.[51] The American mortuary pioneers used an unspecified life-detection apparatus, with the astounding result that out of twelve hundred "corpses," six came back to life. The same article claims that a Dutch waiting mortuary had a similarly high frequency of apparent death: out of one thousand "patients," five revived, thus making the incidence of *Scheintod* 0.5 percent in both populations. This story is likely to have been one of many canards spread by the German pro-mortuary propagandists, who for their agitation were not above using false arguments and invented horror stories about the tortures of live burial. There exists no other record of any waiting mortuary in the United States, before or after. In Britain, appeals to build waiting mortuaries were made at various times, but a combination of parsimony, hygiene, and xenophobia worked against the idea taking root in British soil.[52] The medical profession stood united to condemn the continental fear of premature burial and agreed that these filthy foreign inventions should not be admitted into England's green and pleasant land.

THE ANNALS OF the German waiting mortuaries are a "damned" chapter of history: few people outside Germany know anything about these extraordinary establishments.[53] Even German writers on the subject concentrate on architectural and social aspects and avoid the central questions: Why were these bizarre hospitals for the dead built? Why were they maintained for a period of more than a hundred years? Did they ever serve any purpose? The first question is relatively easy to answer; the other two are more complex. Christoph Wilhelm Hufeland was already a highly respected physician in the 1790s, and respect eventually turned into veneration. Hufeland personified the ideal of a philanthropic, scholarly physician, and many of his ideas gained considerable prominence. To make a proposal to build waiting mortuaries realistic, a strong medical opinion that putrefaction was the only certain sign of death was required. In the 1790s, that was the case in France and in the German states, less so in Holland and Scandinavia, and not at all in Britain. Second, the authors of that proposal had to persuade the authorities to finance the building program; this was possible in many German states because of Hufeland's influence, diplomatic skill, and good relations with clerical leaders. France did not have such an influential figure to head a campaign; it is possible, but by no means certain, that the mortuary system would have prevailed anyway, had the French Revolution not disrupted the plans of Mme Necker and Count von Berchtold. Toward the mid-1800s, Hufeland's stature was such that every time any German doctor or clergyman was bold enough to question the utility of the *Leichenhäuser*, outraged colleagues reminded them that the mortuaries were *Hufelands Erbe*—the legacy of Hufeland—and that any criticism of the mortuaries was the same as questioning the authority of the great man himself. When a certain Dr. Varrentrapp stood up during the International Congress of Hygiene in Brussels in 1852 and said that the *Leichenhäuser* should all be closed down, since there had been no case of resurrection in any of them for many years, the German doctors were furious.[54] Hufeland had quite a few disciples, and well past the mid-nineteenth century the majority of the German medical profession still adhered to the old negativist dogma that putrefaction was the only certain sign of death. Although it was clear, by this time, that no other country was going to follow the German example, and in spite of internal opposition from both the clergy and ordinary people, the larger cities either maintained their old *Leichenhäuser* or actually built new, more advanced ones. This was probably due to a combination of conservatism and medical uncertainty. The German medical profession's acknowledged inability to diagnose death with sufficient certainty had lost for physicians the power over the *Leichenhäuser*, and the authoritarian governments chose to maintain the old system and to enforce the obligatory deposition of corpses in the waiting mortuaries.

How many people's lives were actually saved by the German *Leichenhäuser*? In 1843, the French anti-premature-burial activist Léonce Lenormand claimed that a Berlin apothecary had told him that in two and a half years ten people had revived in the Berlin waiting mortuary,[55] but this story turned out to be a complete invention.[56] A German writer claimed that, by 1847, no apparently dead person had been discovered at any German *Leichenhaus*, and no one brought up evidence to the contrary.[57] The Frenchman Jules-Antoine Josat, who was actually trying to build a case for the construction of waiting mortuaries in France, in 1845 visited the mortuaries of Frankfurt, Sachsenhausen, Mainz, and Munich. He could conclude only that in the preceding twenty-three years these institutions had supervised more than 46,500 corpses without a single detection of apparent death.[58] Nor could another French anti-premature-burial campaigner, active in the 1860s, find any evidence of people's having revived in the German *Leichenhäuser*.[59] In the 1890s, the French anti-premature-burial propagandist B. Gaubert wrote to the Munich and Frankfurt authorities to find out whether any people had returned to life in their mortuary establishments.[60] Herr Ehrhart, the burgomaster of Munich, replied that several people had been resuscitated in the city's *Leichenhaus*, although he did not give any details. Herr Schmitt, the director of the Frankfurt *Leichenhaus*, responded that in 1840 the corpse of a girl had been kept in the mortuary at least nine days, owing to the absence of putrefaction; the vigil had not ended until the corpse burst open—indeed, almost exploded—in front of the grieving parents. If it had been the *Leichenhaus* director's intention to demonstrate the utility of his establishment to his French correspondent, his letter fell badly short of the mark; if this dismal tale was at all representative of what happened within the walls of his establishment, it offered instead a good argument for closing it down. At about the same time, the Englishman

William Tebb made a similar inquiry: the only thing he came up with was the following story. In the Munich *Leichenhaus*, a little girl had once revived, having been discovered sitting on the corpse bed, playing with the white roses used in the floral decorations. When the child's mother arrived at the *Leichenhaus* and saw her child alive, the joy was so overwhelming that she fell dead on the spot.[61] The story does not tell whether her lifeless body was lifted onto the same corpse bed recently vacated by her daughter, but this does not matter; it has been demonstrated that the story was a complete fabrication by some unscrupulous journalist.[62] It is clear that the ultimate decision, in the 1890s, to remove the alarm systems from many waiting mortuaries, rested on the reality that for many years not a single case of apparent death had been detected in any German *Leichenhaus*.[63]

VI

Security Coffins

And there they saw their daughter,
As the moonbeams on her fell,
In her narrow coffin sitting,
Ringing that solemn bell.

—*Mrs. Seba Smith, "The Life-Preserving Coffin,"* Columbian Lady's and Gentleman's Magazine, *January 1844*

As we know, the two greatest objections to the German waiting mortuaries were that they cost a fair amount of money to build and maintain and that many people found them more than a little sinister and unpleasant. As the early enthusiasm for these hospitals for the dead palled, many Germans even went so far as to break the law to keep the bodies of their deceased relatives away from the loathsome corpse beds of the *Leichenhaus*. Some of the German anti-premature-burial propagandists took up their cause and admitted that the mortuaries had their faults: it was an undignified way to treat the dead, exposing them to a humid atmosphere among rows of other putrefying corpses, particularly if not only relatives of the dead but also various paying visitors were strolling about at will. Furthermore, the truly overpowering stench emanating from the larger *Leichenhäuser* was a serious sanitary hazard, they claimed. By the 1790s, an alternative way of safeguarding against the dreaded *Scheintod* had been suggested: the security coffin. The earliest example of an attempt to construct such a coffin involves none less than Duke Ferdinand of Brunswick, the ruler of one of the German states. The duke had read about the horrors of premature burial, and in good time before his death, in 1792, he ordered some skilled carpenters to construct a special coffin. It had a window that let in light, an airhole to prevent suffocation, and a lid with a lock-and-key mechanism instead of being nailed down. The duke had a pair of keys put in a special pocket in the shroud—one for the coffin lid and the other for the vault door.[1]

Security coffins as advanced as the one manufactured at Duke Ferdinand's orders were completely out of reach for ordinary people. It was also a much more difficult task to construct a security coffin that safeguarded against premature burial in an ordinary grave in the churchyard than to do so for burial in a vault. But this did not prevent various enthusiastic inventors from suggesting alternative devices.[2] P. G. Pessler, a German village parson, came up with a novel idea in 1798. He suggested that every time a coffin was buried in the churchyard, a hollow tube should be connected to it. A rope leading to the nearby church bell should be passed down through this tube, so that the prematurely buried individual could make his or her resurrection known by ringing a merry peal of celebration.[3] The parson's plan did not take into account either the weight of the bell or the feebleness of the poor wretch in the narrow box underground. Pastor Beck, a colleague of his, instead recommended that an opening in the coffin be fitted with strings, bells, and an airhole. Another well-meaning German cleric suggested that all buried coffins be fitted with a tube looking rather like a ship's speaking trumpet. The local parson should take a stroll through the churchyard every morning and stop by each recent grave to ascertain, through the sense of smell, whether the putrefaction of the body was sufficiently well advanced to permit the tube to be withdrawn. If there was a lack of putrid odor, the coffin should be disinterred after a few days for inspection by a doctor; if the parson was startled by a cry of "Hilfe!" through the speaking trumpet, the disinterment should of course be performed more expediently.[4]

The inventor Johann Georg Hypelli disagreed with these ideas: the

An undated German plate showing a curious apparatus to save prematurely buried people, not unlike that described by Pastor Pessler. *From the author's collection.*

sick and feeble prematurely buried person did not have the strength to shout through the speaking trumpet or to pull the bellrope; anyway, at night, when the watchman slept, no person was on guard to hear the feeble peal of the frantically ringing bells.[5] He instead proposed that a powerful mechanical alarm bell be put in each larger cemetery and that each coffin be equipped with a special contraption just above the head of the deceased. Any person buried alive would instinctively try to sit up, Hypelli wrote, and the resounding bang as the forehead met the coffin lid would trigger the release of a string that set the bell ringing. The defect of this method was that the coffin had no airhole, but Hypelli was probably an adherent of the theory that the apparently dead could survive, or even wake up in the coffin, without access to fresh air. He stressed the low cost of his invention: just two or three bells would suffice for a large cemetery, fewer than that even among poor people, if they were fitted on high, greased poles to keep people from stealing them. A feeble old man could be given a shed to live in just underneath the bell, thus saving the cost of a proper watchman.

Many ideas like these cropped up in the 1790s, when the concern for the apparently dead was at its height in the German states. It was even suggested that only very light coffins be used and that each buried person have an ax and a spade in the coffin, thereby enabling him or her, it was hoped, to tunnel a way up to rejoin the rest of humanity.[6] Even the kindly Hufeland had to admit that these notions gave more credit to the Christianity and philanthropic zeal of these German parsons than to their intelligence and sense of practicality.

THE FIRST INVENTION of a security coffin that had any serious claim to practicality was that of Adolf Gutsmuth, the city physician in Seehausen, in Altmark. In 1822, he presented a security coffin of his own design that had a long tube linking the buried coffin with the world above. Through this tube, air and light could be admitted into the coffin if the person inside triggered a mechanism. An extra advantage was that it was also possible to administer food and drink to the prematurely buried person by means of another tube inside the larger one, while the coffin was being exhumed. The heroic Dr. Gutsmuth tested his coffin by having himself buried alive in it under five feet of firmly stamped earth. He was dug up after an hour, alive and well thanks to his security system. His next project was even more hazardous. He had himself put into a tightly sealed coffin, in order to find out exactly how long a person could live after being buried alive. The outcome is unfortunately not known, except that Dr. Gutsmuth survived. He was again buried alive in his security coffin later in 1822. This time, he stayed underground for several hours and had a meal of soup, beer, and sausages served through the coffin's feeding tube. He then made a philanthropic speech through a speaking trumpet that had been fitted to the coffin; it is reported that this speech could be clearly heard by a crowd of earnest, well-dressed, respectable spectators standing around the doctor's premature grave.

In 1827, another German practitioner, Dr. von Hesse of Neustrelitz, in Mecklenburg, presented an improved version of Dr. Gutsmuth's coffin.[7] From the lid of the buried coffin, two detachable tubes, 13.5 feet long and 1.5 inches in diameter, extended up to ground level. The one mounted by the head end of the coffin had a speaking trumpet that allowed the prematurely buried person to call for help. It was also possible to ring a bell mounted on top of the tube. Dr. von Hesse tried out his coffin in the same way as Dr. Gutsmuth; he was buried alive in it under five feet of earth, without becoming a martyr to science, but also without attracting as much publicity as had his eccentric fellow practitioner. The Hanoverian doctor Johann Georg Taberger thought that Dr. von Hesse and other inventors had approached the subject too lightly. He criticized the impractical schemes they proposed: a weak, sick individual who had just awakened from a cataleptic trance could hardly pull ropes to set heavy bells in motion, or call for help in a stentorian voice that echoed through the churchyard. Nor could one compare the live burial of a strong, fully conscious man to that of a paralyzed, speechless wretch awakening in a coffin; this individual would have neither the power nor the wits to pull strings, open tubes, or work intricate mechanisms.[8]

Dr. Taberger's own security coffin left nothing to chance. The corpse had strings tied around its head, hands, and feet, and these were connected with a rope that led up through a tube, reaching a bell on top of it. The slightest movement of the individual in the coffin would set the bell in motion. The security coffin had many clever technical details that testified to the Teutonic thoroughness of its inventor. The bell was surrounded by a housing that kept the wind, or birds, from setting it into motion. A fine meshwork inside the tube prevented insects

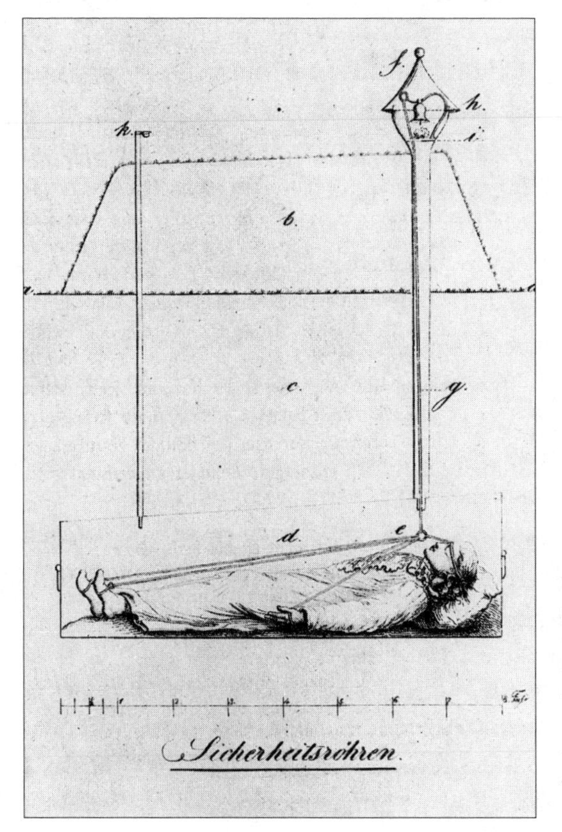

A security coffin described by Dr. Johann Gottfried Taberger in his book *Der Scheintod* (Hanover, 1829). The slightest movement of the hands, feet, or head would ring the bell in the "belfry" above the grave.

from flying or crawling down to annoy the prematurely buried person. A cunning drainage system was to prevent rainwater from dripping down the tube, and thus from adding the Chinese water torture to the horrors of a premature tomb. Most important, the security tube was permanently open to admit air. In a praiseworthy gesture of early environmentalism, Dr. Taberger recommended that a sponge soaked in talcum of chlorine be thrust into the tube, to reduce the efflux of putrefactive fumes from the coffin's "chimney." If the bell was triggered by the prematurely buried person, a second security tube could be inserted at the foot end of the coffin. By the use of a powerful bellows, air circulation, which was thought to be very beneficial for further restoring the dormant life of its luckless inhabitant, could be achieved inside the coffin. Dr. Taberger hoped that a collection of money among philanthropic Germans would enable all vergers to purchase a supply of his security tubes, which were cheap and reusable. The only drawback, which he could not in any way remedy, was that a watchman had to be on duty day and night in each cemetery; otherwise the frenzied ringing of the poor wretch underground would have little effect.

The 1820s saw an attempt to combine the idea of the waiting mortuary with that of the security coffin: the so-called portable death chamber.[9] This was a small *Leichenhaus* built for just one inhabitant, with a hermetically sealed door with a window. A string led from the corpse's hand to a bell on top of the death chamber. The plan was to employ a corps of gravediggers and watchmen in each cemetery; they would go on regular rounds to inspect the corpses through the windows of the death chambers that emerged on top of each grave, and be ready around the clock to resuscitate each "corpse" that rang the bell. If there were clear signs of putrefaction, a trapdoor in the bottom of the death chamber would be triggered, and the corpse would fall down into a previously dug grave underneath it. A second door could then be inserted in lieu of the coffin lid. A contraption of this sort was actually in regular use in Eberstädt, in Gotha. A report in the civic archives of Weimar, the site of the first *Leichenhaus* in Germany, criticized it severely: a paralyzed, speechless victim of the *Scheintod*, who did not even have the power to ring the bell, would be disposed of in the most cruel fashion and dumped into a premature grave by the same hands that had been employed to save him or her.

IN THE SECOND half of the nineteenth century, the German obsession with security coffins continued, and upwards of thirty different versions were patented.[10] A certain Richard Strauss, from Schweidnitz, in Silesia, constructed several security coffins of increasingly advanced design. One had a filter to keep sand and air from falling into the face of the prematurely buried person, and a patent lamp that could be lit inside the coffin. Another had a complete air-conditioning system, with a powerful pump and a filter through which the air was recirculated. The crowning touch was that the bell was replaced with a powerful firecracker, by whose detonation the prematurely buried person could make his or her resurrection known even in a large and desolate cemetery. The products of other imaginative German inventors included a mechanical brass hammer that smashed a glass pane in the coffin lid, a pyrotechnical rocket that was launched through the security tube of the coffin, and a loud siren mounted inside the tube. This latter variant also had a similarly penetrating siren in the other end of the tube, which could be used as a last resort, its blasts echoing in the narrow coffin. If the corpse did not respond to this final wake-up call, the tube was to be withdrawn.

Britons who wanted to insure themselves against being buried in a state of apparent death could order coffins equipped with the Bateson Life Revival Device, an iron bell mounted in a miniature campanile on the lid of the casket, the bellrope connected to the presumed corpse's hands through a hole in the coffin lid. Bateson's Belfry, as it was called, was of course most useful before burial, but those wealthy and cautious enough could have the apparatus buried with them. The Victorian gentleman may have hoped that, should he revive in his padded coffin, the sinister underground knelling from the Belfry would call some atten-

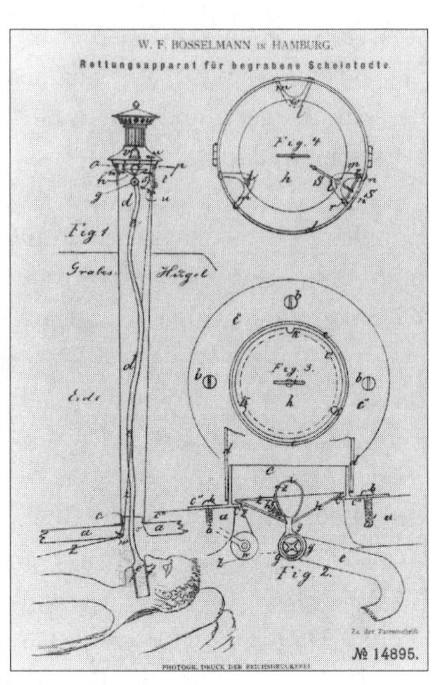

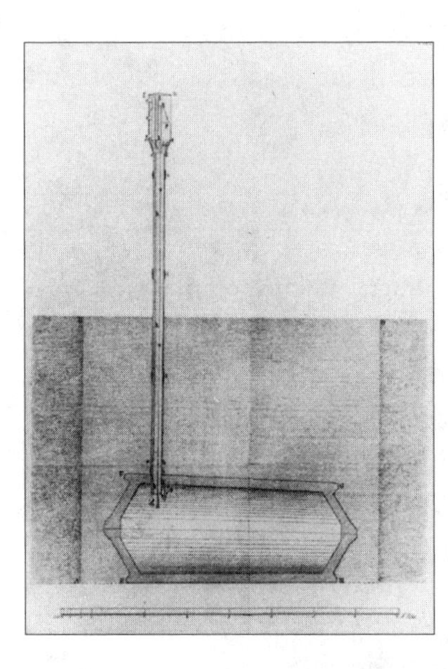

Two drawings of another German security coffin. *From the author's collection.*

tive verger or gravedigger to the rescue, and that the coffin would be unearthed and the creaking lid opened by a polite servant, with the words "You rang, Sir?" George Bateson, whose Belfry was patented in 1852, was a successful inventor and mechanic. He described his security device as "a most economical, ingenious, and trustworthy mechanism, superior to any other method, and promoting peace of mind amongst the bereaved of all stations of life." Quite a few security devices were sold, and Bateson was awarded the Order of the British Empire by Queen Victoria for his services to the dead. According to some accounts, Bateson later became more and more preoccupied with the horrors of premature burial. He built increasingly complicated alarm systems for his own coffin. Not trusting even his own apparatus, he later rewrote his will, instead asking to be cremated. Finally, the wretched man became quite insane and feared that his directions would not be followed; in a fit of desperation, he doused himself with linseed oil and set himself ablaze in his workshop, preferring a premature cremation to the lingering risk of premature burial.[11] A later British invention was a security coffin with a lid fitted with powerful springs, for use in vault burials.[12] There also exist drawings for a coffin with an ejection seat, but it does not seem to have been manufactured.

In the United States, at least twenty-two patents for security coffins were applied for between 1868 and 1925, some of them by optimistic Germans who wanted a worldwide market for their coffins, but the majority by American inventors.[13] The earliest application was filed in August 1868 by Franz Vester of Newark, New Jersey, a German-American who had brought his enthusiasm for saving apparently dead people with him to his new country. His security coffin remains unique in the annals of these inventions. The coffin lid had a sliding door, to which was fitted a detachable, hollow cylinder with a ladder bolted to it.

A drawing for a coffin with "safety springs" that made the lid swing open if the mechanism was triggered by its inhabitant. *From the* Burial Reformer *magazine of October–December 1911.*

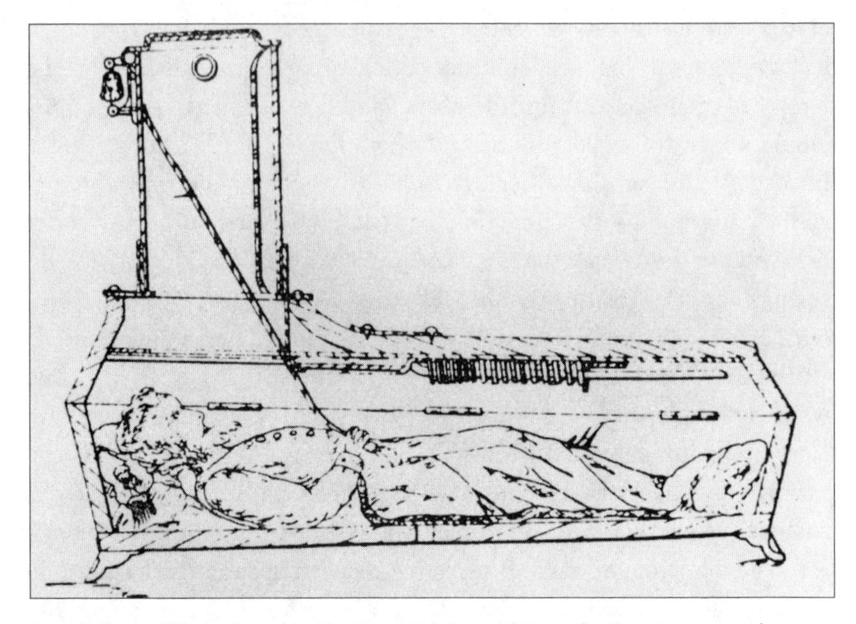

A security coffin designed by Mr. Franz Vester, of Newark, New Jersey, and patented on August 25, 1868. On awakening, the prematurely buried person was supposed to ring the bell by means of the rope. If this failed to produce results, he should climb up through the square tube attached to the coffin lid, by means of the ladder. The inventor had clearly not counted on his fellow Americans' inclination toward stoutness; most of them would have got stuck when attempting such a maneuver. *From the article by H. Dittrick in the* Journal of the History of Medicine *3 (1948): 161–71, reproduced with permission.*

Upon recovery, the apparently dead person would presumably crawl through the opening, climb up the ladder, open another door, and thus reenter the world of light. If, on the other hand, he got stuck halfway, he did not have to await the enforced slimming cure needed to reduce his girth; he could ring a bell mounted outside the tube to call for assistance. The coffin contained a box of food and wine, which he could enjoy while ringing for the sexton. The detachable tube could be removed and the sliding door closed if, on later inspection, life was again proven to be extinct. Following the proud tradition of Dr. Gutsmuth and Dr. von Hesse, Franz Vester decided to try his coffin out himself, although his residence in the United States had diluted his philanthropy toward prematurely buried fellow humans and instead produced a healthy dose of financial astuteness. It is recorded that five hundred paying New Yorkers saw one of his shows and that Vester stepped into his coffin "with more alacrity than could be expected of a man about to submit to so *grave* an experiment." A little girl put a flower wreath on his head before the coffin lid was screwed down, and a brass band played a mournful tune as the coffin was lowered into a six-foot-deep grave. After an hour, Vester was dug up, and the crowd rushed forward to embrace and congratulate him; the punning newspaper reporter quoted earlier added that Vester did not at all appear *decomposed* by his sojourn underground.[14] The later American patent coffins became increasingly advanced, featuring various electrical signals triggering flags, bells, and rotating lights. The most elaborate ones even had an electric light, a heater, and a telephone. There were untrue rumors that Mary Baker Eddy had purchased a coffin equipped with a telephone, and later been buried in it, but at least one of her fellow

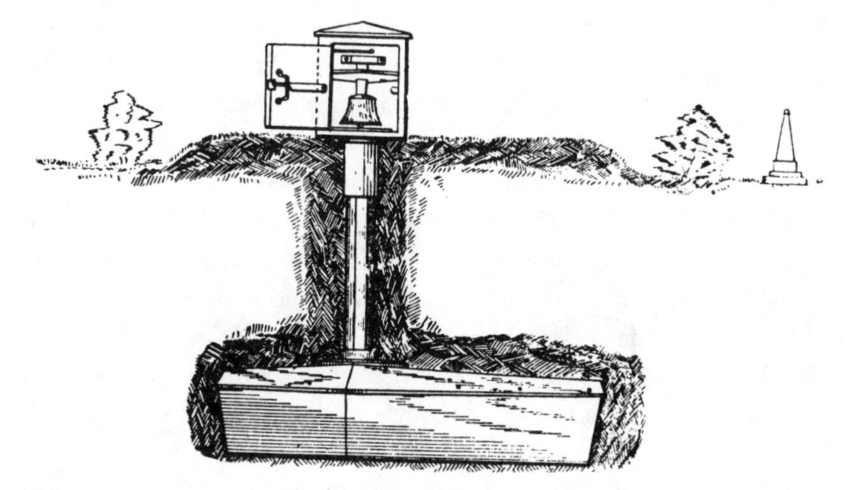

A security coffin with a bell and an airhole, designed by Mr. August Lindquist, of Charlton, Iowa, and patented on June 20, 1893. *From the article by H. Dittrick in the* Journal of the History of Medicine *3 (1948): 161–71, reproduced with permission.*

Americans became a customer. In 1908, a wealthy Lousiana lady named Mrs. Pennord, who greatly feared a live burial, was buried in a security vault equipped with airholes, and a telephone connected to the cemetery keeper's house was put on her bedside table.[15]

COUNT MICHEL DE KARNICE-KARNICKI, a chamberlain to Czar Nicholas II, and a doctor of law at the University of Louvain, was a distinguished Russian nobleman who spent much of his time in France and Italy. His family came from Warsaw, but he had independent financial means and could travel about Europe as he wanted. The count's carefree life was changed, however, by a horrific experience while he was attending the funeral of a young Belgian girl. When the first shovels of earth were thrown upon her coffin, she awoke from her death trance, and her piteous screams haunted him ever since. He decided to invent a mechanical apparatus to save the lives of those prematurely buried. His first invention was a coffin with a large glass pane in the lid, which could be opened or smashed by the prematurely buried person. This security coffin was of course only useful for vault burial. But the count's labors continued: he wanted to invent a security coffin suitable for burial in the earth, which he intended to make available to rich and poor alike. After several years of hard work, Count de Karnice-Karnicki was able to present a particularly advanced security coffin, eponymously called Le Karnice, before an international audience at the Sorbonne in 1897. Beforehand, he had patented it both in Europe and in the United States. The idea was that every time someone was buried, a long tube three and a half inches in diameter was to be connected with an aperture in the coffin, the end of it emerging above ground rather like the periscope of a submarine. A visit to one of the German waiting mortuaries had given Count de Karnice-Karnicki a healthy respect for the foul miasma that the putrefying corpse exuded, and a metallic box attached to the top of the tube was closed to the outside world, to keep the gases from the coffin from escaping. The slightest movement of the chest or arms of the "corpse" in its coffin would set off an intricate mechanism, which would open the tube to admit air into the coffin. At the same time, a bell would ring and a flag wave on top of the mechanism, which bore some resemblance to an American mailbox. This would, it was hoped, call the diligent vergers and gravediggers to the rescue. Apparently, Le Karnice worked quite well when tried out at the Sorbonne before a large gathering of foreign ambassadors, medical men, hygienists, and gentlemen of the press, many of whom praised the philanthropic zeal of the Russian nobleman. In the *Journal d'Hygiène*, several authors agreed that Count de Karnice-Karnicki's security coffin was superior to any such earlier contraption, and that it should be widely used to prevent premature burials.[16] The influential Parisian

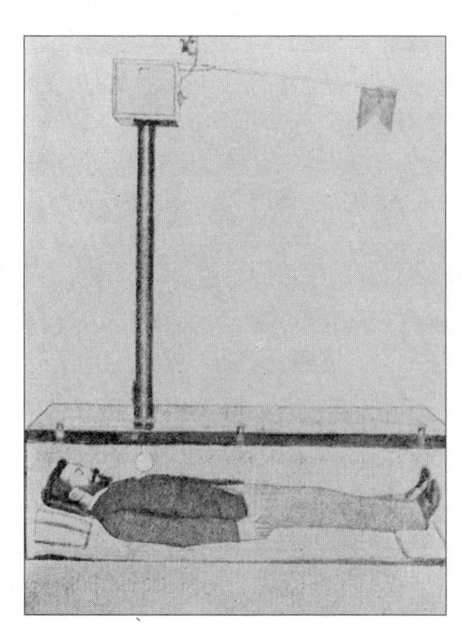

"Le Karnice," a before-and-after sketch of Count de Karnice-Karnicki's invention being triggered by a man buried alive. *From the author's collection.*

hygienist Professor Charles Richet told the press that the problem of apparent death was no more and that lethargy had been vanquished by Le Karnice.

Count de Karnice-Karnicki then went on a road show through France and Belgium to demonstrate his security coffin. The initial reactions to his invention were positive, but one of the later demonstrations was not a success. The count had arranged for an assistant to be buried alive in a coffin fitted with Le Karnice; this individual was to trigger the mechanism after spending a few minutes underground. But the flag did not wave and the bell did not ring, and the audience grew increasingly nervous; consternation could be observed on the faces of Count de Karnice-Karnicki and his associates. They grabbed their shovels and started digging frenziedly. Fortunately, it turned out that the airhole had been functional, and the intrepid assistant could climb out of the coffin unaided. Some journalists were present at the ill-fated demonstration, and the faint aura of ridicule that had plagued the count's invention from the start now erupted in ribald articles in the Paris boulevard newspapers. Some of the experts who had previously praised Le Karnice now recanted, admitting that a security coffin that worked *only sometimes* was not a very valuable invention.

This failure would have crushed a less determined man, but Count de Karnice-Karnicki continued his crusade undaunted. He held several further demonstrations of the coffin and badgered numerous medical societies and academies for their approval. The count's French agent, M. Emile Camis, wrote to the newspapers to say that those individuals who had tried to belittle Le Karnice should be ashamed of themselves for their mean and spiteful attacks on the magnanimous, philanthropic Russian. He claimed that Count de Karnice-Karnicki had no fewer than four times been able to prevent the burial of people "stricken with lethargy" and that this accomplishment, if nothing else, should give credit to his claim that a mechanical apparatus to prevent premature burials was urgently needed. The road show went on to Berlin, Brussels, London, and Milan. When it came to Turin, the count arranged a bizarre demonstration. A seventy-eight-year-old man named Faroppo Lorenzo had consented to be buried alive in Le Karnice. Buried on December 17, 1898, he was not disinterred again until December 26. Signor Lorenzo's only comment, when dug up after this dreary underground Christmas vacation, was that it had been damned smelly down there! The prefect of Milan and the inspector of cemeteries were so impressed by this demonstration that they awarded the count a signed certificate that his coffin was fully functional.[17]

To argue his case further, Count de Karnice-Karnicki wrote a pamphlet entitled *Vie ou mort*.[18] The count's prose was rather bombastic and florid: for example, he wrote that premature burial was such a torture that it made all the torments that the Romans inflicted on the Christians at the Circus in Rome seem mild by comparison. He deplored the flippant journalists who had made fun of him, and blasted the conniving funeral directors who had promised to order coffins but failed to make good their promises. Worst of all were Dr. Laborde and Dr. Vallin, two French hygienists who had presented a scathing report on Le Karnice to the Académie de Medecine in Paris. They had argued that the mechanism was too sensitive and could easily be triggered by putrefactive changes, but the count emphasized that even the most feeble, paralyzed creature should be able to maneuver the coffin. Moving to statistics, he claimed, on the authority of the American physician Dr. Carleton Simon, that 1 person out of 30, 000 buried was still alive when put into the earth. He contrasted this dubious figure to another one, which he had found in a French newspaper: only 1 out of 11,000,000 railway passengers was killed. Surely, it was more dangerous to be buried than to travel by train, and it was thus "a monstrous inconsistency" of the French government to improve railway safety, while scoffing and deriding those earnest, uncomplaining philanthropists who cared for the apparently dead!

In December 1899, Le Karnice was taken to the United States by the enterprising M. Camis, who gave a lecture on its merits before the Medico-Legal Society of New York.[19] The American medical profession was so impressed with the count's invention that a special committee of the Medico-Legal Society of New York, chaired by Dr. H. Gerald Chapin, reported that there was ample cause to fear that premature burials occurred frequently and that nothing but the systematic use of Le Karnice could prevent them. The Detroit physician Dr. Noah E. Aronstam, who had been present when Le Karnice was examined by the Medico-Legal Society, also wrote an article to recommend its use.[20] For several years, one of the prototypes of Le Karnice was exhibited in a showroom at 835 Broadway, but the American funeral directors were not particularly impressed with it. Probably, the word had spread from Europe that the mechanism was much too sensitive, that it did not allow for even the slightest movement of the corpse during decomposition.

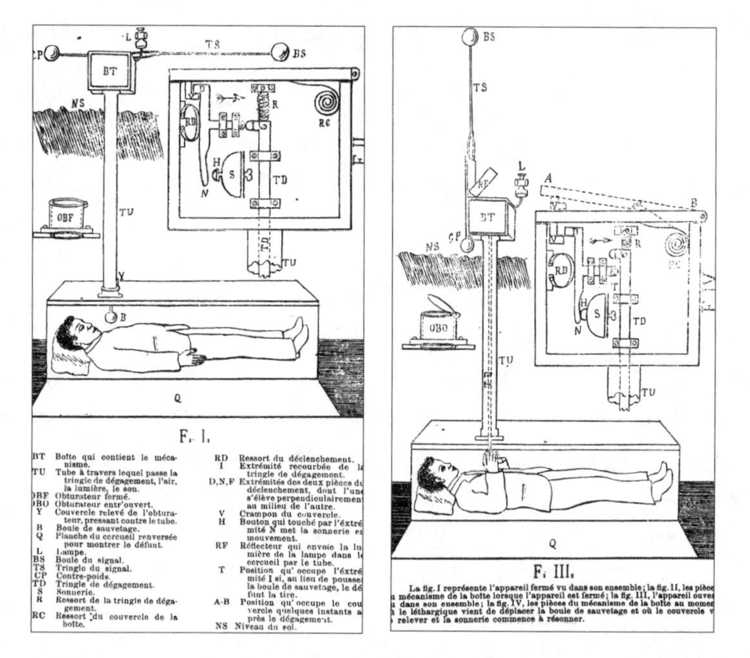

Build your own Le Karnice: a very detailed sketch from a French advertisement pamphlet for Count de Karnice-Karnicki's invention. *From the author's collection.*

WE TODAY KNOW much more about human physiology than the early nineteenth-century inventors did, and this allows us to point out some glaring defects in the construction of the nineteenth-century security coffins. First, a person enclosed in a normal-sized, airtight coffin, the lid of which had been screwed down, would perish within sixty minutes because of the lack of oxygen. The security coffins that lacked a permanent air supply would thus fulfill their purpose only if the apparently dead individual regained consciousness within an hour, which was probably not the intention of their inventors. For any person whose state of *Scheintod* lasted longer than sixty minutes, the security device would certainly prevent the disagreeable experience of awakening in a coffin, but at the price of killing the "client" from suffocation. The security coffin designed by Dr. Taberger, and the majority of the late nineteenth-century ones, had a permanent air supply, and would thus allow their apparently dead inhabitants to survive much longer. But Dr. Taberger's coffin had another serious defect. We now know that the putrefactive changes of a corpse are accompanied by swelling of the abdomen, as a result of gas formation within the abdomen, and also by contractures of the arms and legs of a variable degree. This process would certainly set off the mechanism of the coffin, whose bell strings were actually tied to the corpse's head, hands, and feet. Le Karnice had a similar defect, its sensitive mechanism being positioned where it could easily be triggered by the putrefactive changes. Both Dr. Taberger and Count de Karnice-Karnicki held that the mechanism had to be triggered by the slightest movement, conjuring up a vision of horror as the paralyzed victim of the *Scheintod* desperately tried to move a finger to call for assistance, but this excessive concern instead made their coffins practically useless. The false alarms resulting from putrefactive changes of the corpses buried in coffins equipped with these forms of security devices would have led to many distasteful and distressing scenes in the cemeteries, with bells ringing and little flags waving on the "mailboxes," had either of them been used to any extent.

The optimal security coffin would thus have a permanent air supply and an alarm mechanism that demanded intelligent effort, like the turning of a handle. The later models, with electrical wiring to a bell in the cemetery office were particularly useful, since they disposed of the necessity of having a watchman patrolling the cemetery day and night. But despite all the rhetoric of Dr. Gutsmuth, Dr. Taberger, Count de Karnice-Karnicki, and other inventors of security coffins, the idea never really caught on. The parsimonious and utilitarian elements stood united to condemn such needless expenditure, and the supporters of the *Leichenhaus* movement viewed the various attempts to invent security coffins as a betrayal of Hufeland's noble and philanthropic idea. Le Karnice was marketed in many countries, but in spite of the determined efforts of Count de Karnice-Karnicki, few coffins were actually sold. The ultimate reason for the failure of the security coffins to go into serial production must be sought in the field of psychology rather than in that of mechanics. The intended customers for the security coffins, people fearful of premature burial who wanted a safeguard at any price, were concerned not with the alarms that would sound if the mechanism was triggered by putrefactive changes but rather with the risk of the coffin's mechanism not operating properly, thus consigning its luckless denizen to a horrible death. After all, a person could stay alive for a long time in a coffin with an air supply, screaming and frantically pulling the faulty mechanism in a desperate attempt to communicate with the world above. What would happen if the coffin's mechanism jammed, or the strings came off in the trembling hand that pulled them? What horrors would ensue if some mischievous youth stole the "mailbox" above earth, or if the cemetery watchman decided to take a nap instead of patrolling his beat, leaving the bell to ring and the flag to wave jerkily until finally, they came to a halt as the apparently dead person, seized with despair, tore his hair and frantically grappled with the unyielding walls of the narrow prison?

VII

The Signs of Death

Our very hopes belied our fears,
Our fears our hopes belied;
We thought her dying when she slept,
And sleeping when she died.
—*Thomas Hood,* The Death-bed

The acceptance of putrefaction as the sole reliable sign of death, as signified by the construction of waiting mortuaries, represented the extreme defeatist standpoint in the nineteenth-century debate about apparent death and the risk of premature burial. The German doctors effectively confessed that they were unable to tell the living from the dead in the absence of bodily decomposition. In contrast, the positivist response, in this time of growing confidence and knowledge in medical science, was to try to discover novel, more reliable signs of death. A useful sign of death should be very specific, and no living people should test positive—otherwise they would be buried alive. It should also be highly accurate, and none, or at least very few, corpses should test negative—otherwise multitudes of putrefying corpses would be kept above ground for long periods of time, leading to risks of contagion as well as ridicule for the medical attendants involved. A very important point was that the sign of death in question be applicable not only by specialists but also by medical practitioners of modest knowledge and dim intelligence; to propose a test of life or death that could be applied by ignorant laypeople was considered too dangerous by far. As we know, Antoine Louis had already tried to propose certain eye changes, like softening of the eye bulb and drying out of the cornea, as perfectly reliable signs of death.[1] The contemporary medical profession was incredulous, rightly so, since the changes proposed were neither marked nor specific enough. The standardized application of this test by ignorant country practitioners could have led to a full-scale disaster, in which living people were buried wholesale. Nor were the brutal, old-fashioned tests of life, as described in Winslow's thesis, of great practical value; the resuscitated patient was unlikely to be particularly grateful to his or her physician, if the latter cut off all ten fingers in the efforts to stimulate the spark of life, or thrust a red-hot poker up the rear passage. There is a case report about an Italian carabineer named Luigi Vittori who "died" from an asthma attack, but was revived by having his nose severely burned with a lighted taper: for the rest of his days, the scarred, crimson beacon in the middle of his face told of his narrow brush with death.[2] There was no shortage of similarly crude tests of life: strong smelling salts and pinches of foul-smelling snuff should be put into the corpse's nostrils, fanfares blown on a powerful bugle close to the unprotected ear, and the body's most sensitive areas rubbed with stiff, prickly brushes.[3] A Swedish writer recommended that a crawling insect be put into the corpse's ear.[4]

Antoine Louis had also proposed another method of testing life, or at least stimulating the vital spark in the apparently dead person: with a powerful bellows, he administered an enema of tobacco smoke. One of the pipes of this remarkable apparatus was thrust into the anus of the apparently dead person; the other was connected, by way of a powerful bellows, to a large furnace full of tobacco.[5] Such enemas of tobacco smoke were thought to be very beneficial and were used to try to revive not only people presumed dead but also drowned or unconscious individuals. In 1784, the Belgian physician P. J. B. Previnaire was given a prize by the Academy of Sciences in Brussels for a book on

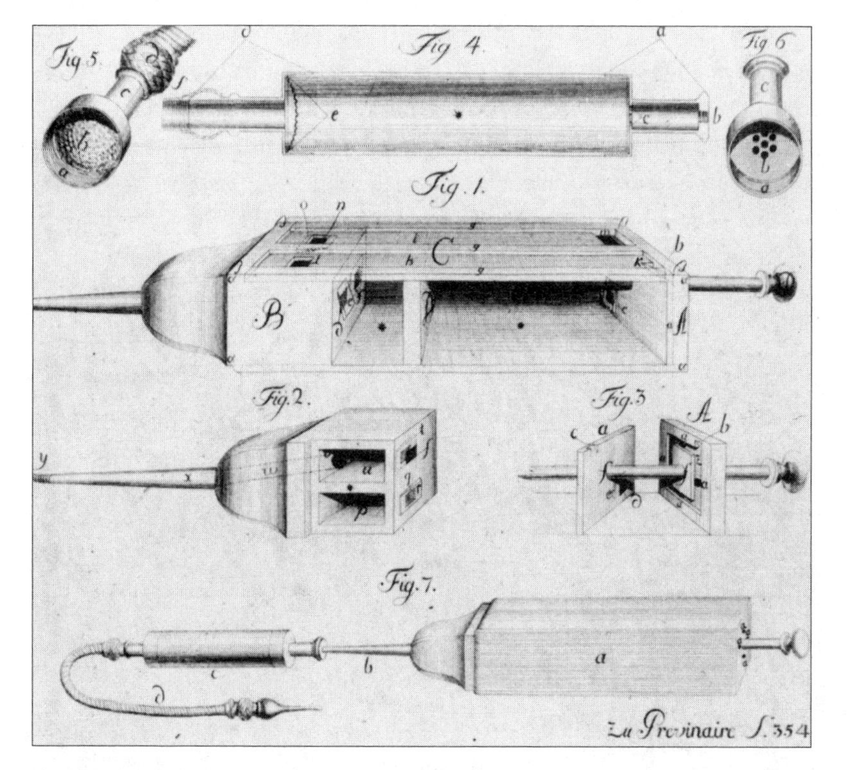

The fearful-looking *Doppelbläser*, an apparatus for administering enemas to test the viability of corpses, described in Dr. P. J. B. Previnaire's *Abhandlung über die verschiedenen Arten des Scheintodes* (Leipzig, 1790).

apparent death, which described and depicted an improved bellows for enemas of tobacco smoke, which he called *Der Doppelbläser*.[6] These enemas were regularly used well into the nineteenth century, particularly in Holland; modern science has discerned no physiological rationale for their use, except that the pain and indignity of having a blunt instrument violently thrust up one's rear passage must have had some restorative effect.[7] Another eighteenth-century apparatus for testing life was based on the directly opposite principle. It had been observed that the sphincters of the intestinal tract relaxed after death and that this would facilitate the passage of air through the digestive tract.[8] An eccentric inventor suggested that a nozzle should be forced down the corpse's throat, and a powerful air pump applied by two strong men; the grotesque scene that must have followed can be imagined by those familiar with dead bodies, but remains mercifully hidden to laymen.

Already in the 1790s, there appeared some alternative proposals of useful tests of life or death; the reader will, at this stage, not be sur-

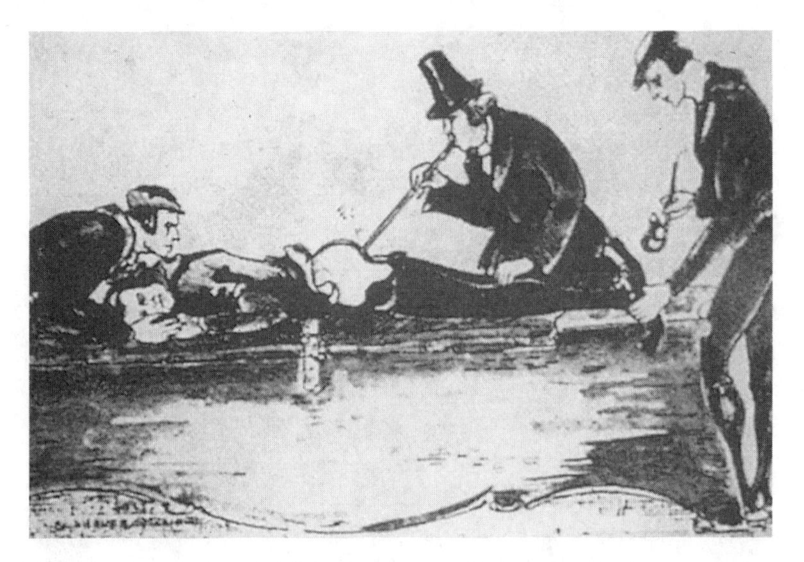

A brave German doctor administers an enema of tobacco smoke to a corpse in this curious late eighteenth-century plate. *From the author's collection.*

prised that they involved German doctors and inventors. The use of electricity in resuscitation was attracting considerable interest at this time.[9] In 1796, the German Dr. Caspar Creve described his *Metallreiz*, a galvanic apparatus to test the irritability of a muscle.[10] In this procedure, the biceps muscle of the corpse's arm was laid bare and the apparatus brought in close contact with it; if the individual was still alive, a twitch would be produced. There is no reason why this procedure should not have worked, provided the electric current was strong enough, as Creve claimed that it was, after a series of experiments on animals and on recently dead people. Doubts about the practicality of his invention were widespread, however, and when another enthusiast tried the *Metallreiz*, he could not get it to work.[11] In 1808, Hufeland wrote that there had been no further research on Creve's *Metallreiz*, because of questions about its practical applicability.[12] Galvanism was still considered dangerous and uncertain, and bereaved family members were unlikely to appreciate that the doctor wanted to cut up their dear departed for his ungodly experiments.

In 1805, Dr. Christian August Struwe proposed another electrical apparatus for use in determining whether a person was alive or dead.[13]

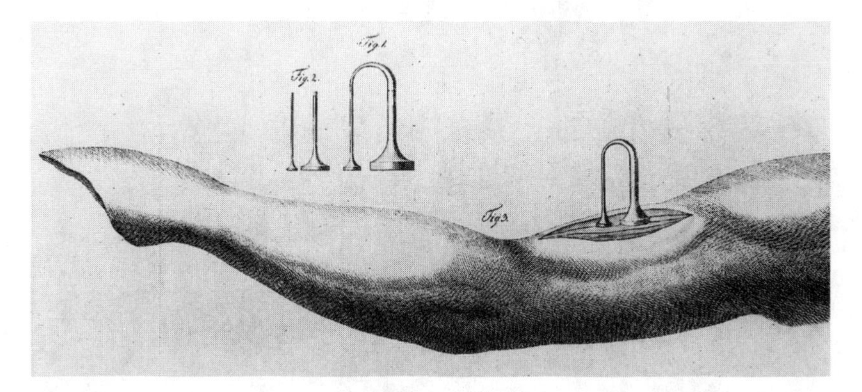

Dr. Creve's *Metallreiz*, a galvanic apparatus for making certain that the "corpse" had really died: a muscle was dissected out on the corpse's arm, and the galvanic plate was applied to make it twitch if the individual was still alive.

He called it the *Lebensprüfer*—the Tester of Life—boasting that it was applicable on land, at sea, in wartime; its use all over the world would save thousands of people from the most fearful of extremities, that of being buried alive. The *Lebensprüfer* consisted of a wooden staff with two strings full of copper and zinc cones. The metal cones were connected through pieces of linen soaked with a sal ammoniac solution, to generate an electric current, and each string ended with a conductor. When one of these conductors was pressed against the eye, and the other against the upper lip, a twitch of the eye and mouth muscles should occur if the person was still alive. Again, the concept is a reasonably sound one, and Dr. Struwe had performed experiments similar

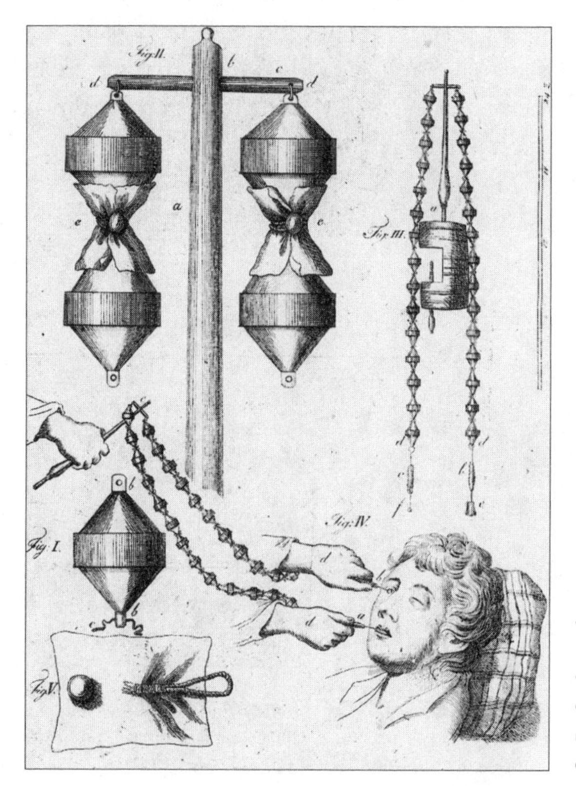

A plate from Dr. Christian August Struwe's *Der Lebensprüfer* (Hanover, 1805), showing how a corpse is tested for lingering signs of life by means of the author's electrical apparatus.

to those of Creve, but a lack of faith in electrical machinery and the high cost of this relatively complicated device conspired to deprive this invention of whatever practical importance it might have had.

BY THE 1830S, there had been little further worthwhile German research concerning the signs of death, since Hufeland's doctrine of the *Scheintod* and putrefaction as the only reliable sign of death had become generally accepted among medical scientists. Entrenched in their foul-smelling *Leichenhäuser*, the German physicians dared not suggest that the great man had been wrong and that there really was a reliable sign of death apart from putrefaction. Nor was there much progress in England, where the medical establishment was wholly complacent, viewing the Continental preoccupation with premature burial with a mixture of amusement and disgust. Thus, any significant initiatives with regard to the signs of death would have to come from the French-speaking part of Europe. The influential physiologist Xavier Bichat had considerably furthered knowledge about the process of death by his experimental studies.[14] Bichat used vivisection as his main experimental approach: he dissected or burned dogs, kittens, and other animals alive, or studied their struggles when slowly drowning. Death could be instantaneous, like that of in a guillotined criminal; it could be a drawn-out ordeal, a succession of events in various tissues of the body, in cases of lingering disease or in an animal that he was slowly torturing to death. It was clear to Bichat that in an animal strangulated to death, black (deoxygenated) blood permeated the tissues of the brain and destroyed its "animal activity"; the organic life of the heart and lungs later perished in separate events. Bichat defined life as the composite of the forces throughout the body that resisted the influence of death; thus no organ could be called the seat of life. His opinions therefore formed the logical extreme of the gradual process of decentralization of death that had been set in motion by Bruhier's questioning the central role of the heartbeat and respiration in the determination of life. Bichat's influence on the physiological study of life and death was beneficial, but his extreme vitalist standpoint could also be interpreted to imply that no sign of death, particularly one connected with a single organ like the heart, could be relied upon. The influential forensic scientists François Emmanuel Fodéré and Mathieu Orfila drew a logical consequence of Bichat's theories: there was no question of a single sign of death being the supreme one; rather, an entire array of tests of life or death should be performed, to monitor the state of different organs. They were confident that the signs of death taken together were absolutely certain, although each of them alone was inconclusive.[15]

From 1800 until 1835, more than thirty French medical theses dealt with the signs of death. These dissertations could be divided into three equally large groups.[16] One defended the certainty of the signs of death along the lines of Louis and others, often implying that it was an insult to modern medicine to accuse any doctor of not being able to tell a living person from a corpse. A second group admitted that most signs of death were fallible, but not putrefaction and cadaveric rigidity. A third claimed that all signs of death were fallible except putrefaction. Some of the writers in the third group went so far as to demand waiting mortuaries in the French cities. In 1837, Professor Pietro Manni, a toxicologist active at the universities of Naples and Rome, donated 1,500 gold francs to the Academy of Sciences in Paris, for use as a prize for the best work on the signs of death and the means of preventing premature burials.[17] This valuable donation had the desired effect of further stimulating French research into the signs of death and their reliability. The first contest for the Prix Manni, in 1839, had quite a few entrants. The Academy of Sciences rightly upheld its strict criteria, however, and considered none of the contributions worthy of the prize; the result when the Prix Manni was re-advertised in 1842 was equally disappointing.

In 1846, the Prix Manni was advertised for a third time. One of the contestants was the thirty-one-year-old physician Eugène Bouchut. He had been greatly impressed by René Laënnec's invention of the stethoscope, and the subsequent demonstration of the clinical usefulness of this instrument in diagnosing diseases of the heart and lungs.[18] Bouchut suggested that this instrument be used to diagnose the cessation of heartbeat. He was of the firm opinion that if the stethoscope was properly employed to determine the definitive lack of heartbeats, for a period of two minutes, no doubt should remain, and the dead body could be buried without further delay. He supported this view by a large number of experiments on animals killed or sedated in different ways and by many observations of human beings in extremis. Condemning the waiting mortuaries as needless and macabre, Bouchut considered his countrymen's fear of being prematurely buried greatly exaggerated; it was fueled by the many inaccurate and sensation-seeking pamphlets and newspaper reports on this subject. Besides Bouchut, quite a few others competed for the Prix Manni. There was no lack of cranky ideas about how to tell corpses from living people. The suggestions included pouring astringents into the stomach by means of a stomach pump and putting multitudes of leeches near the corpse's anus. The German Dr. Christian Friedrich Nasse invented what he called the *Thanatometer*, a thermometer introduced into the stomach in a long tube, which was supposed to measure whether the temperature of the core of the body was low enough to permit a diagnosis of death.[19] His Austrian colleague Dr. Johann Nepomuk Hickmann invented an electrical apparatus for testing life, which he claimed to be superior to Struwe's *Lebensprüfer*, although the principle appeared to be the same.[20] The French doctor Jules-Antoine Josat invented a pair of strong pincers, with which to pinch the presumed corpse's nipples,[21] and the Englishman Barnett recommended that the arm of the corpse be scalded with boiling water; a blister would appear if the individual was still alive. Another odd notion, proposed by a German named Middeldorph, was that a long needle, with a flag attached to one end, should be thrust into the heart of the apparently dead individual; the flag would be waving merrily if the heart was still beating.[22]

In 1848, Dr. Rayer of the Academy of Sciences could announce that Eugène Bouchut was the winner of the Prix Manni.[23] This prestigious award attracted international publicity, and Bouchut's position was further strengthened by the fact that both Rayer himself and some other leading academians, including the neurobiologist François Magendie and the anatomist Etienne Serres, gave him their active support. It is interesting to speculate whether the pivotal decision of the academy was wholly prompted by confidence in Bouchut's discovery, or whether it also wanted to obturate the mouths of the raucous anti-premature-burial activists, who enjoyed the backing of the newspapers, and asserted many untruths in their loathsome pamphlets. By preferring Bouchut's paper to the mediocre, negativist suggestions that waiting mortuaries be built everywhere, the academy certainly made its views on the subject clear. The following year, Bouchut published his investigations as a book, which also contained an acerbic attack on the French anti-premature-burial alarmists.[24] He praised Louis and Bichat as important forerunners and criticized Bruhier's books as sensationalist and unreliable. Unlike the earlier writers, he saw no point in wasting time quoting older opinions on the absence of an audible heartbeat as a sign of death; one could not blame men like Celsus, Pliny, Bruhier, Winslow, and von Haller for not having available the admirable work of Laënnec. The publication of Bouchut's *Traité des signes de la mort* attracted much interest throughout Europe, becoming the most influential and constructive book on this subject since the beginning of the nineteenth century. Some modern historians have found the award of the Prix Manni to Bouchut amusing—the notion that listening to the heartbeat is a good indicator of death appears very natural to modern people.[25] But that was by no means the case in the 1840s, and Bouchut was in fact very brave to suggest this novel sign of death. After all, a good death sign should be applicable also by an elderly, somewhat deaf country practitioner. If Bouchut's sign proved fallible, a multitude of living people without an audible heartbeat would be buried alive, and he himself would become a marked man in French medicine. Laënnec's invention of the stethoscope was still quite recent, and some conservative clinicians were not entirely convinced of its diagnostic value. Also, the stethoscopes of that time were fairly primitive: they were made of solid wood and resembled crude hearing trumpets. When I tried one out not long ago, the contrast to a modern instrument was striking. In fact, it is not unreasonable to claim that Bouchut was wrong, that a living person can lack an audible heartbeat, particularly when one is limited to using a mid-nineteenth-century stethoscope. Conditions like pulmonary emphysema, a large pleural effusion, or an exsudative pericarditis can serve to isolate the transmission of the heart tones to the chest. More important in clinical practice, the slow, weak heart action of a person intoxicated with strong sedatives is not always audible with the stethoscope. But Bouchut's suggestion was specific and accurate enough to be of great practical value, and it had the advantage of simplicity.

Another widespread misconception concerning Bouchut's discovery is that the publication of his book virtually ended the controversy about apparent death and the risk of premature burial that had raged throughout Europe since the publication, a hundred years earlier, of the works of Bruhier.[26] In fact, the German medical establishment remained obdurate, and the waiting mortuaries were to stand for another half century. A certain Professor Plugge, of Darmstadt, objected that if Bouchut's ideas were taken seriously, a multitude of newborn infants in a state of asphyxia would be wantonly consigned to an unnecessary grave.[27] A reviewer in an English journal was wholly skeptical concerning the utility of Bouchut's discovery.[28] In his native France, too, Bouchut faced a torrent of criticism.[29] Jules-Antoine Josat, the inventor of the nipple pincher, claimed that he had once, in a cholera epidemic, encountered a living person with no audible heartbeat, and several of the other unlucky contestants for the Prix Manni also made unkind remarks about the unreliability of Bouchut's discovery and the mass carnage that would occur if it was taken seriously. Numerous leading clinicians, from hospitals all over France, sent to the Academy of Sciences lists of living people with no audible heartbeat. According to one account, from an anti-premature-burial activist, the cowardly Bouchut was overwhelmed by the massive opposition in the early 1850s and was willing to agree that his sign of death was fallible.[30] Although this claim is not supported by less biased sources, it is likely that Bouchut was not a little affected by the many distinguished clinicians opposing him; had he not been supported by the distinguished forensic specialists Alphonse Devergie and Auguste-Ambroise Tardieu, his future would have looked uncertain indeed. Bouchut tried to reach a compromise by increasing the period of auscultation from two to five minutes. This compromise was not accepted by his opponents, and the debate still raged in the 1860s. In particular, the French anti-premature-burial activists regarded Bouchut as a traitor and lost no opportunity to publicize various cases where medical men had not been able to hear the heartbeat in still-living people.

Nor was there any shortage of new proposals of signs of death or tests of life. In 1861, a certain Dr. Plouviez suggested that "acupuncture of the heart" by means of a steel needle be performed in difficult cases, where Bouchut's test had failed, and several commentators agreed.[31] The surgeon Jules-Germain Cloquet and the physiologist Jean-Vincent Laborde instead proposed that a shiny needle be plunged into the bulk of the biceps muscle.[32] If the person was alive, the needle would become metabolized and rust, they claimed, but this test strikes me as far-fetched as well as nonspecific. Dr. Marteno, of Cordova, recommended that a candle flame be put half an inch from the finger or toe; if the person was dead, an air blister would form.[33] Again, this test would not have the desired power of discrimination, since such a blister would also form on a recently dead person. In 1879, the Italian Ugo Magnus proposed that a string be tied tightly around one of the person's fingers; if he or she was still alive, the finger would become blue and swollen as the venous efflux of blood was stopped by the ligature.[34] This test is not as good as it might seem, however, since persons in extremis might have very pronounced peripheral vasoconstriction, which would lead to their being falsely declared dead. In 1874, the Italian Dr. Angelo Monteverdi proposed a better idea.[35] He could prove that a subcutaneous injection of liquid ammonia produced an inflammatory reaction in a living person, but not in a corpse. This sound test of life depended only on the ability of the doctor in charge to perform the injection correctly—hypodermic syringes were a novelty at this time, which limited the applicability of this test considerably.

A certain Dr. Léon Collongues proposed a singular test. If the finger of a living person was stuck into the doctor's ear, the involuntary slight muscle movements produced a buzzing sound; this sound was absent if the finger of a corpse was inserted in a similar manner.[36] He boldly claimed that this test was superior to those of Louis and Bouchut. Dr. Collongues also believed that the nature of the buzzing sound from the patient's finger in the doctor's ear was an important diagnostic aid in various diseases. Similarly cranky was another suggestion from the aforementioned Dr. Laborde: the tongue of the deceased should be rythmically pulled for a period of three hours.[37] He had previously asserted, with some right, that pulling the tongue was advisable when performing artificial respiration, but this was because it helped clear the airways and not because the maneuver possessed some strange stimulating influence on the central nervous system. Dr. Laborde claimed that he had once been called to an unconscious, asphyxiated woman sitting in a dentist's chair, after a dangerous mishap with the anesthesia. As the distraught dentist and anesthetist were looking on, he grabbed her tongue with a strong forceps and started pulling it violently. The outcome was a happy one, and the woman was saved, although she complained that the pain in her tongue was so great that she had forgotten all about both her period of asphyxia in the dentist's chair and the toothache that had caused the consultation in the first place! Dr. Laborde's "Physiological Treatment of Death" also worked in animals: an unconscious cow and a swooned English bulldog had both been resuscitated by the intrepid doctor. He had invented a tongue-pulling machine for use in the mortuary: it was driven by a crank, turned by an unskilled mortuary assistant. The assistant complained of his dreary and unprepossessing duties, which had to be kept up for three hours before the presumed corpse could be consigned to the grave. Dr. Laborde solved this problem by replacing him with an electric tongue-pulling machine, photographs of which he proudly reproduced in one of his books. A reviewer in a religious magazine was so impressed with this machine that he wrote that this valuable French invention deserved to become known, and vigorously applied, all over the world.[38]

IN 1867, A French nobleman, the marquis d'Ourches, donated 5,000 francs to the Academy of Sciences in Paris for a prize to be awarded for the discovery of a scientific method of recognizing with certainty the signs of actual death. A much higher sum, 20,000 francs, should be awarded to the person who proposed a reliable sign of death that could be determined by an uneducated layperson as well. The marquis was an eccentric character, who had dabbled in spiritualism and somnambulism. He had also spent much time taming animals, hoping that one day, dogs, cats, rabbits, tigers, and men could be educated to live together in peace and harmony. His Prix d'Ourches was well advertised in the newspapers when it was thrown open in 1868, and a record number of 102 suggestions were forwarded to the academy, not counting 12 that arrived too late. Most of the candidates were natives of France, but some hailed from Germany, India, and the United States. Interestingly, the medical men were in a minority; the claimants included several inventors, clergymen (priests, pastors, and rabbis), retired military officers, several women, a grocer, and a hairdresser. The last individual stated that he knew a certain manner of judging from the appearance of the hair whether a person was dead or alive; he did not go into details about its application, but offered to do so after he was awarded the prize. The academy did not take up his offer. The German inventors, of course, brought forth various security coffins and signaling systems for mortuaries. Another German offered to hypnotize the presumed corpses to find out if they were dead. Other suggestions included injecting strychnine into the stomach, shining a powerful searchlight into the eyes to provoke a contraction of the pupil, and burning the temples or the region of the heart with a white-hot iron. One contestant suggested that a wrench, a shovel, and a ladder be put into the coffin to facilitate an escape; another, that a trumpet be put into the mouth of the presumed corpse, so that an awakening in the coffin would be heralded by a ghostly fanfare from below ground. An English medical journal was outraged at these silly ideas, particularly that laypeople were allowed to compete for the prize: "What serious work can be looked for in this motley assembly? What lucubrations could emanate from the brains of these persons ignorant of every scientific fact, who have the pretension to unveil the mysteries of life and death. . . . "[39] The French forensic specialist Alphonse Devergie also ridiculed the marquis and his prize, but for other reasons: "Suppose for a moment that a sign of death obvious to the whole world would be discovered. What would happen? People would not call a physician to verify death. . . . Let us therefore abandon the unfortunate idea that the signs of death must be vulgarized."[40]

The prize board of the Prix d'Ourches was busy for several years drawing up a shortlist, and it took even longer before the winners could be announced. To the disappointment of the hairdresser and grocer, the winner was Professor Weber, a forensic specialist from Leipzig, who had suggested that a hard brush be used to brush some part of the skin of the dead body some hours after death. If the individual was parchment-like in texture after the brushing had taken place, death was certain. Awarded the smaller prize of 5,000 francs, Weber would have had a claim to the great prize of 20,000 francs, had some of the commissioners not objected that they had tried to repeat the professor's procedures on some corpses and had seldom seen the promised result. Some of the other contestants received honorable mentions for observations on body temperature and various eye changes after death; a certain M. de Cordue was awarded 500 francs for his observations on the effect of the flame of a candle upon the pulp of the finger.[41] The great prize was returned to the heirs of the marquis d'Ourches, who probably welcomed this outcome, since they had been left destitute by the eccentric nobleman's will and lived in the poor quarter of Saint Germain.

Another philanthropic Frenchman, M. Dusgate, was inspired by the great publicity surrounding the Prix d'Ourches to donate 2,500 francs to the Academy of Sciences for a quinquennial prize to be awarded to the discoverer of a reliable sign of death or a way of preventing premature burials.[42] The first essays were received in 1880, but no prize was awarded until 1890, when the prize board made the remarkable decision to award it to a certain Dr. Maze, who had proposed that putrefaction was the only certain sign of death and that all dead bodies be supervised in waiting mortuaries and cremated after signs of putrefaction had set in. This decision was opposed by the French medical profession: by this time, the technical development of the stethoscope and the confidence of the doctors in using it had improved considerably. When the triumphant Eugène Bouchut published the third edition of his book, in 1883, nothing remained of the doubts he had confessed to after the onslaught in the 1850s, nor were there any objections to his theories from the French medical profession, except for some cranks and enthusiasts in the anti-premature-burial lobby. Bouchut firmly asserted that the absence of an audible heartbeat was a certain sign of death, that the agitation of the French anti-premature burial activists rested on unreliable newspaper evidence, and that the German waiting mortuaries were unnecessary.[43] By this time, he had an acknowledged position in his own country and an appreciative international audience.

In 1895, one of the competitors for the Prix Dusgate was Séverin Icard, a physician at the Grande-Miséricorde children's hospital in Marseilles. His interest in premature burial was one of long standing, and some recent distressing episodes in his practice had prompted him to try to invent a safeguard.[44] In 1893, Dr. Icard had been called to examine the body of a lady in Marseilles; he declared her dead, but the relatives were extremely fearful of premature burial and wished him to come once more, since they worried that she might still be alive. Dr. Icard returned and offered to apply the test of Middeldorph to set their minds at rest. With no further ado, he flourished a long, shining needle and plunged it into the corpse's heart. Her husband and children gave a cry of horror: "Yes, now she is dead—*But you have killed her, doctor!"* Dr. Icard did not dare to venture into that part of town after nightfall for some time and silenced the press only with difficulty. The following year, Dr. Icard was again consulted in a difficult matter of life or death. The eighteen-year-old Zéphirine Maniel had apparently died and was declared dead by the family doctor. But her parents were not convinced, since there was no cadaveric rigidity. They appealed to the mayor, who ordered Icard and four other doctors to settle the matter. The distraught parents stood by the bed crying, "Zéphirine est vivante!"—"Zéphirine is still alive"— and a large and hostile mob gathered outside the house. The five doctors did not hesitate to declare the girl dead, and the mayor ordered that the funeral proceed as planned. By this time, however, fifteen hundred people had assembled, and the police had to protect both the doctors and the funeral procession from their wrath. There was even an attempt to grab and break open the coffin, to make sure poor Zéphirine was not buried alive. The next morning, all the newspapers proclaimed that a dreadful error had been made by the (named) doctors and that another premature burial had taken place.

After several years of research, Dr. Icard had come up with a quite useful test for life or death.[45] By subcutaneous injection of a solution of a strongly fluorescent compound, it could be determined whether the individual still had a functioning circulation of the blood: in that case, the skin turned a vivid yellow and the eyes emerald green, and this phenomenon lasted until the compound had been metabolized. Dr. Icard had perfected this method by injecting both dead and living dogs with his fluorescent solution. The living animals had been lent to him by trusting colleagues, whose reaction when the yellow, green-eyed dogs were returned to them has unfortunately not been recorded in Dr. Icard's voluminous writings. Séverin Icard injected many species of animals, even hibernating hedgehogs and tortoises, with the desired result. In dead animals, or in human corpses, no reaction took place. In 1895, and again in 1900, he received the Prix Dusgate for his valuable invention. Dr. Icard also suggested an original test for putrefaction: a piece of paper with the words "I am really dead," written in acetate of lead was folded and put down the nose of the "corpse"; the reaction with the dioxide of sulfur in the putrefactive gases made the words stand out vividly. Whereas the fluorescein test is perfectly rational, provided the injection is properly done, Dr. Icard's other idea did not have the same specificity: there are many diseased conditions, like suppurating tonsillitis and severe dental caries, that may lead to the production of sulfur dioxide. Nor is the production of putrefactive gases always enough to cause the changes desired. When an English doctor tried to repeat Dr. Icard's experiment, he saw a definite change in only one out of six corpses.[46]

VIII

Skeptical Physiologists and Raving Spiritualists

To die is natural; but the living death
Of those who waken into consciousness
Though for a moment only, ay, or less
To find a coffin stifling their last breath,
Surpasses every horror underneath
The sun of Heaven, and should surely check
Haste in the living to remove the wreck
Of what was just before, the soul's fair sheath.

How many have been smothered in their shroud!
How many have sustained this awful woe!
Humanity should shudder could we know
How many cried to God in anguish loud,
Accusing those whose haste a wrong had wrought
Beyond the worst that ever devil thought.

—Percy Russell, "Premature Burial," from Burial Reformer *magazine, 1 (1906): 33*

By the early nineteenth century, the danger of premature burial had become one of the most-feared perils of everyday life, and a torrent of pamphlets and academic theses were dedicated to this subject by writers all over Europe. In almost every country, a literature on this gruesome topic was readily available, ranging from the solemn medical thesis and the philanthropic call for more waiting mortuaries to pamphlets written by fanatics who claimed that more than one-tenth of humanity was buried alive, and to horror-mongering compilations of bugaboo stories like that of the notorious Köppen. The British nation had managed to resist the Continental obsession with premature burial

A Swiss lady awakes in her coffin just as it is to be lowered into the grave; her relatives run away, believing that she is a ghost. *A drawing in the author's collection, marked "Basler Hinkender Bote, 1822."*

in the late eighteenth century, but in the 1810s the French pamphlets invaded the British Isles, provoking a good deal of magazine writing on the subject. Moreover, the British reading public did not have to rely wholly on the importation, from Germany and France, of alternately silly or sadistic stories on prematurely buried people being saved from the tomb or dying amid hideous torments. In 1816, the book *The Danger of Premature Interment*, by Joseph Taylor, of Newington Butts, appeared in London.[1] This imaginative Englishman had previously written on apparitions and ghosts, "Canine gratitude," and "Anecdotes of Remarkable Insects"; his diligence in seeking odd reports of apparently dead people's coming to life was commendable, more so than his judgment in commenting on them. The original Lady with the Ring tale was of course reproduced in his collection, as well as a rather amusing English variation on the same theme, said to have involved Sir Hugh Ackland, of Devonshire. This worthy had died from a fever, it seemed, and two footmen sat up with the corpse, which was laid out in a coffin. One of them, remembering that his late master "dearly loved brandy when he was alive," poured a glass of brandy down the corpse's throat. Sir Hugh immediately responded to treatment and came sputtering to life; he survived for several years, and the impudent footman received a handsome annuity. Three years after Taylor's book was published, the British public was again reminded of the Continental debate about premature burial. In a book on various aspects of death, the Reverend Walter Whiter claimed that people were regularly buried alive by mistake, not only in Germany and France but also in the heart of Britain.[2] He had read the English edition of Bruhier, and he also recounted a tale of a German lady who woke up from her attack of *Scheintod* "with a most pitiable shriek" as the lid of her coffin was to be nailed down. Whiter believed that the warm earth possessed remarkable powers to restore life; indeed, he recommended that earth baths be employed as a means of revivification in cases of death trance. This state of affairs only further increased the danger of awakening in the tomb, Whiter wrote: "The warm genial Earth possesses, I believe, mighty virtues for assisting the Resuscitative process, and I grievously fear, that the examples of revival in the Grave are more frequent than the World, amidst all their alarms existing on Premature Interment, has yet ventured to conceive."

If Taylor's and Whiter's books were relatively balanced, John Snart's *Thesaurus of Horror; or, The Charnel-House Explored* was probably the most ludicrous and gruesome book ever to appear in its particular literary genre.[3] Snart was a writer on mathematics, about whom little is known except that he was *very* frightened of being buried alive. The nature of the book's contents can be easily gleaned from the following deplorable poem, introducing its main theme:

But if the fertilizing earth restore
The dubious fragment of a borrow'd life,
Can man's most desp'rate scuffle force the grave,
Or must he, grappling, bathe himself in blood,
And burst his eyeballs in the vain attempt!!!

Snart claimed, "Death, in a natural way, is the common lot of all mortals, from which none can escape; but this horrid supplement of supernumerary woe, is a second death, extrinsic of that originally intended and far surpassing it in misery!" He also detailed the tragic fate of Baron Hornstein, a popular courtier in Bavaria, who had been buried with great ceremony after his sudden demise, from apoplexy, at an early age:

> Two days after the Funeral, the workmen entered the Mausoleum, when they witnessed an object which petrified them with HORROR!!! At the door of the sepulchre lay a body covered with *Blood*! It was the mortal remains of the *favourite of Courts and Princes*! The Baron had been BURIED ALIVE!!! On recovering from his Trance, he had forced the lock of his Coffin, and endeavored to escape from the Charnel-house! It was impossible! He therefore, in a

> fit of desperation, had dashed his brains out against the wall!!! The Royal Family, and indeed the whole city, are plunged in grief at the horrid catastrophe.

To dash one's brains out against a wall, a gesture much favored by desperate villains in French novels, would require neck muscles of a prize bull and a skull the thickness of a china bowl; such features would have made the baron an interesting spectacle already when alive.

John Snart had read about what he called the Turkish test of death, which involved blowing air down the corpse's throat by means of a hog's bladder with a tube attached to its orifice. The idea was that the relaxation of the sphincters after death would render it easier for the air to pass through the digestive tract. The sight, smell, and sound of this ludicrous procedure must have been disgusting in the extreme. Snart was convinced that this method was infallible, however, and recommended it for general use to prevent premature burials. The British Library's copy of Snart's book has a letter by Neariah Snart, the author's daughter, pasted into it. This letter contains Miss Snart's solemn promise, before God and Jesus Christ, that her father would never to be buried alive and that she would make sure that the Turkish test was carried out, with a satisfactory result, before he was put into his coffin. John Snart was anti-Catholic, but he also despised the Jews and their religion; his book is full of violent assaults on the Jews for their deleterious habit of early burials. In drawing a parallel to the barbarous practice of burying people before they are dead, Snart looked to the animal kingdom:

> *If asp, or basilisk, or noxious weed,*
> *(Distilling mortal venom through its frame,)*
> *The more they're fatal,—sudden, more their aid*
> *To snatch me from this horrid fate I crave!*
> *Tarantulas and rattle-snakes attend,*

And hard-mouth'd alligators, come and dine!
For, though ye paralyze and rend the frame
Ye have not learnt to bury men alive!!!

Snart's passionate description of the horrors of the untimely tomb deserves to be quoted at length, as an interesting expression of the deepest and most unreasonable fear of a live burial:

> Behold the hapless victim of this horrid custom, upon the return of life, shut in the clay cold prison!—he lifts! ah, no!—his trembling hands to procure him that relief he feels so much the *need of*; and though before grown *feeble* by *disease*, made desperate now, by the maddening sense of his hapless situation and *lost estate*! But yet the attempt is stopped!—the coffin is shut, shut for ever! screwed down!—loaded with unrelenting earth! Terror,—despair,—horror, torments, unknown beliefs, seize on him! Madness,—rage,—all! all!—no power to live! no power to die! No power, alas, to cry for aid! but pent, barricaded, and pressed by accumulating condensation! The brain distracted! the eyes starting from their sockets! the lungs ruptured! the heart rent asunder by unnatural impulses! the ducts and glands suffused, the emunctories choked by surcharge of feces, rendered viscid by incalescence and external resistance; and every vein and artery bursting in the super-human conflict! The office of inosculation (baffled) tries in vain to force its valves and runs retrograde, bathes the poor grappling victim in extravasated blood *without*, and forms new channels *within*, in this dreadful scuffle, which knows no cessation or abatement, till coagulation's influence stagnates and deprives him of all thought, and he becomes a fermentable mass of *murdered, senseless, decomposing matter*!!![4]

With such alarmist pamphlets on the market, it is not surprising that some English people, usually from the upper classes of society, left legacies to their family physicians to protect themselves against this gruesome fate. The antiquary Francis Douce left two hundred guineas to the surgeon Sir Anthony Carlisle to see to that his heart was taken out after death, and his friend Mr. Kerrick did the same, with the added safeguard that his own son was to supervise the procedure. Mr. Ritson, another antiquary, wished his coffin to be filled with quicklime to render resurrection impossible. There are quite a few examples of Britons taking similar precautions, ranging from the sensible—to delay burial for three or four days after death—to the most gruesome mutilations, worthier of a butcher's yard than a mortuary. Lady Dryden, of Northamptonshire, left an eminent physician fifty pounds to make sure her throat was cut before burial, and the doctor performed this task. Mrs. Elizabeth Thomas, of Islington, directed that her physician pierce her heart with a long metal pin, and again the loyal doctor performed this postmortem operation. The writer Harriet Martineau left her doctor ten guineas to see to it that her head was amputated. Mr. William Shackwell, of Plymouth, in Devon, requested that all his fingers and toes be amputated, to ensure that he was really dead. A more frequent, and much more sensible, precaution was to request that the jugular vein be cut before burial, as did the poet Edmund Yates and the actress Ada Cavendish, among many others. Bishop Berkeley, the novelist Lord Lytton, and the poet Daniel O'Connell, also dreaded premature burial, and put precautions in their wills to escape this fate, and Lady Burton, wife of the explorer Sir Richard, ordered in her will that after death her heart be pierced with a needle and her body then dissected and embalmed.[5]

Still, the popular fear of premature burial never reached the same heights in Britain as it did in Germany and France, except during cholera epidemics, particularly the one in 1831–32. In January 1832, there were claims that a laborer in Haddington had been buried alive; the wretched man had swept the grave clothes away from him, but the brutal attendants had nailed down the coffin lid anyway. On investigation, the doctor in charge said that the man had probably had a muscular spasm postmortem, as was sometimes observed in cholera cases. In

Sunderland, preposterous rumors of people being buried alive were also widely spread. The ignorant populace believed that the doctors in charge were getting rid of troublesome patients by drugging them with laudanum and sending them off to the churchyard in a coffin, or that they were so fearful of disease that they deliberately killed their patients in order not to catch cholera themselves. There was even a rumor that the doctors were actually paid for each corpse they could deliver to the churchyards and that some unscrupulous medical men added to this income by murdering a few patients by having them buried alive.[6] The same distrust of authorities is reflected in the following newspaper account, headed "Singular Scene at Loughborough." An old woman had died at the local workhouse, but some children thought they could hear her groan in her grave. A crowd of garrulous workhouse women arrived, and they believed they heard groans and felt the earth shake from the prematurely buried woman's desperate efforts to escape her confinement. They started digging frenziedly with whatever implements they could find. More than two thousand people gathered, and one of the women stood on a tombstone to deliver a harangue on the wickedness of the workhouse officers, who had deliberately buried the old woman alive to get rid of her. A furious chimney sweep's wife put her fist in the sexton's face and "loaded him with almost every epithet in the genteel vocabulary." But when the coffin was finally opened, the corpse was found to be undisturbed, ice cold, and very dead.[7] It is instructive that these incidents involve distrust of authorities rather than actual fear of an *accidental* live burial.

IN THE 1820S AND 1830s, when the enthusiasm for anti-premature-burial philanthropy was beginning to wane in Germany, the concern for the helpless apparently dead people, and the fear of premature burial, gathered new strength in France. On December 13, 1834, M. Hyacinthe Le Guern read a memorial before the Chamber of Deputies in Paris, demanding that waiting mortuaries be erected in every major city. The deputies took no action. M. Le Guern then published a horrid pamphlet entitled *Rosoline, ou les mystères de la tombe,* allegedly based on a true story.[8] A young lady, Rosoline d'Ab——, described as young, beautiful, and embodying every female virtue, died and was duly buried. Two months later, her heartbroken fiancé was desperate to see her one final time. Deaf to the argument that her corpse was unlikely to be a particularly attractive sight, he clandestinely went into the graveyard in the evening and dug up Rosoline's coffin with the help of an aged domestic. He was horrorstruck when he saw that the coffin lid was shattered and that a fleshless hand had been thrust through the gash in the lid—Rosoline had been buried alive! In such a fearful extremity, one would have expected an excitable young Frenchman to go mad, tear his hair, and embrace the coffin, shaking his convulsively clenched fist toward the sky and cursing God in a shrill voice that reverberated from the nearby tombstones. But Rosoline's fiancé did nothing of the sort. He went home, drank a glass of toddy, and wrote a long letter to Rosoline's physician, calmly stating that the careless medical man had, by his fatal error, condemned his young patient to the most horrible death that could be imagined. The magnanimous young man forgave the doctor, but urged him to take better care in the future and not bury any more of his patients alive. He ended his lengthy missive by recounting a number of famous historical cases of premature burial, quoted from the works of Bruhier and other authorities. M. Le Guern was a stalwart propagandist, active throughout the 1840s with his distasteful pamphlets, although he was never able to surpass his first lugubrious masterpiece. He did not lack readers among his countrymen, and in 1844 he published the sixth edition of his *Danger des inhumations précipitées.*[9] He was seconded by Mme Hortense Du Fay, a lady philanthropist and busybody, who wrote pamphlets against immorality, drunkenness, and the danger of Gypsies and vagabonds stealing children. She had strong views on the subject and claimed that she had once herself been falsely diagnosed as dead by a physician. Her solution to the problem was that all corpses be kept above ground for seven or eight days before burial and that, as an extra safeguard, their arms, legs, and heads be chopped off before they were put in their coffins.[10]

Hyacinthe Le Guern's appeal in 1834 caught the notice of the physician and chemist Jean-Sébastien-Eugène Julia de Fontenelle. They met, and Julia de Fontenelle was treated to a private reading of *Rosoline*. Whether the dreadful pathos of this tale moved him to complete his own treatise on the subject of apparent death and premature burial is not known, but it did not take him long to publish his *Recherches médico-légales sur l'incertitude des signes de la mort*.[11] Julia de Fontenelle was a respected physician and a knowledgeable scientist, though one given to flights of fancy, and his book was to become the most influential French work of its kind for quite a number of years. But this was due more to Julia de Fontenelle's reputation than to any merits his work possessed. In fact, there was little difference, in outline or in reasoning, between his book and those written by Bruhier almost a hundred years earlier. Indeed, Julia de Fontenelle gave Bruhier a handsome éloge for his "judicious observations," which had by no means been refuted by Louis's attack. He fully agreed with Bruhier that putrefaction was the only certain sign of death. The fifty-seven cases of premature burial or narrow escapes in his book contained many old favorites: Archbishop Géron, Cardinal Andreas, the Two Young Lovers, the Lady with the Ring, and several incarnations of the Careless Anatomist. The modern cases added little that was credible. A certain Dr. Fossati told the story of a man buried in a vault underneath a church in Italy. An ominous murmuring noise emanated from the vault, but when the churchgoers told the priests in charge, the learned clerics replied that it was probably the soul of the dead man saying its prayers! After two days, it was decided to open the vault; the spectacle they encountered was so gruesome that the doctor who had wrongly declared the man dead himself died of grief shortly thereafter. Another modern case, of a soldier who had "died" in the 1832 cholera epidemic, but revived in the coffin and gave a cry as it was lowered into the grave, would have claims of credibility, had it not been vouched for by the notoriously unreliable M. Le Guern, the author of *Rosoline*.[12] In 1833, Julia de Fontenelle had traveled in Germany to study the waiting mortuaries there, and his suggested solution to the problem of premature burials was to start an ambitious building program for these hospitals for the dead all over France.

In France, Julia de Fontenelle's book was praised by reviewers as a work of both science and philanthropy. But in Britain, it received a harsh welcome. In a scathing review, the London physician Robert Ferguson made fun of the French nation's collective fear of premature burial.[13] He wrote that in France many people, especially women, were so haunted by this dreadful fate that they almost feared to go to sleep, lest they should wake up with a coffin for their bed and six feet of earth for their covering. Far from trying to calm these exaggerated fears, French physicians like Julia de Fontenelle actually produced books and pamphlets that helped deepen the horror of the grave. In Britain, these terrorists boasted no educated disciples, and the popular fear of premature burial had not caught on as in Germany and France. Much of the blame for the French preoccupation with premature burial rested with the unscrupulous newspaper editors, who multiplied the cases of alleged premature burial to an extent that would have been frightful, had the cases not been refuted by their very number. Dr. Ferguson wrote that while an English country editor in want of a paragraph might invent a tale of a bird of passage being shot out of season, or an apple tree blossoming in October, the French neighbors demanded stronger and spicier tales, and premature burial was a perennial favorite with editors and readers alike. Julia de Fontenelle had been musing that the number of Frenchmen annually buried alive was so great that the military power of his country had to suffer, but Dr. Ferguson was unconvinced. Making fun of Julia de Fontenelle's attempts to recruit the population from the graveyards, he wrote that if the practice of premature burial had been as common as the French author had surmised, the priest's "Requiescant in pace" during evening mass would have been answered by a loud chorus of "Amen" from the surrounding graves.

Turning to Julia de Fontenelle's cases, Dr. Ferguson found much to criticize. Enough of a scholar to see through the old tales of the Two Young Lovers and the Lady with the Ring, he wrote that if the Frenchman had been as well read in English literature, he could have added Romeo and Juliet to his list of pleasant stories. Another case involved a French officer who woke up from a death trance on the dissection table; he made his resurrection known with the words "I perceive the action has been hot!" Equally ludicrous was the tale of an apparently dead abbé, who was carried to the churchyard in his coffin. For some reason, his cat had been put in the coffin with him, and the animal made such a caterwauling in there that the pallbearers stopped to see what was happening. When the lid was opened, out jumped the cat, closely followed by the abbé, who ran off as fast as his legs could carry him. Dr. Ferguson wrote that even a person of moderate intellect would immediately perceive that both these stories had been intended as jokes rather than as medical case reports.

Dr. Ferguson's demolition of Julia de Fontenelle's arguments is not only amusing reading but also highlights the contrasting attitudes toward apparent death and premature burial in Germany, France, and Britain in the 1830s and 1840s. As we saw, the Germans were by this time beginning to doubt the value of their precious *Leichenhäuser*: the bulk of medical opinion still favored the construction of such hospitals for the dead, but both the clergy and the educated middle classes no longer shared this view, to say nothing of the common people. In France, which had no waiting mortuaries, there was on the other hand widespread popular agitation for these safeguards against premature burial, and many people were convinced that premature burials were common and that every one was at risk. Indeed, the intensive production of French pamphlets and medical works on this subject resembled that of Germany in the 1790s. In Britain, the books of Joseph Taylor and John Snart had entirely failed to whip up a similar mass agitation for burial reform, and the medical profession shared Dr. Ferguson's attitude of amused incredulity at these distasteful foreign excesses. In the early nineteenth century, the British people were more concerned with another problem concerning their dead (or maybe not so dead) fellow beings: the practice of robbing graves to steal corpses for the anatomy schools. Instead of inventing security coffins that safeguarded against a premature burial, British inventors patented steel or concrete coverings for the tombs that grave robbers could not penetate. Among the other devices used to ward off the body snatchers were spiked railings around the tombs, or even craftily hidden traps or buried spring guns. If a German philanthropist visiting Britain would have been alerted by stifled cries for help emanating from a recently filled-in grave, he might have learned the hard way that concerns for the dead take different forms in different cultures. His frantically digging hand would have been caught in the jaws of the hidden steel trap or torn by the blast from a powerful shotgun.

IN THE 1840S, the French preoccupation with apparent death and premature burial continued unabated. The production of unwholesome books and pamphlets on this subject exceeded all precedents, and the situation resembled that in Germany after Hufeland's appeal in 1790. Many people put clauses in their wills to safeguard themselves against an awakening in the coffin. The famous composer Frédéric Chopin feared premature burial and before his death in 1849 requested in his will that his body be dissected after death. His fellow composer Giacomo Meyerbeer and King Leopold I of Belgium made similar requests. The irony that many Germans had now got tired of their precious *Leichenhäuser* was lost on the earnest French propagandists, whose expansive plans for waiting mortuaries in every town and major village would have upset the civic finances considerably. In 1843, the country practitioner Dr. Léonce Lenormand published a pamphlet to support the erection of a waiting mortuary in his hometown of Macon.[14] He excelled in detailed descriptions of the torments of awakening inside the coffin, interspersed with the familiar tales of the Lady with the Ring, the Two Young Lovers, and the Careless Anatomist. The latter bogeyman appeared in a novel guise: he started to dissect a Belgian soldier whom he presumed to be dead, but the soldier revived and attacked him fiercely. The Anatomist prevailed after a furious fight, when the soldier fainted from loss of blood.[15] Another horrific tale, similarly lacking in verification, featured a nobleman from Florence, Prince L**, who was declared dead after a lethargic illness. His body was later found lying on the floor of the family vault, where he had died of hunger: "he had awakened from the clutches of death only to find another death, one that was a thousand times more horrible." Lenormand's amazing cases including one avowedly taken from a scientific journal published in London. In 1831, a young Englishman died of typhus fever in France and was buried in a common grave in the churchyard. On the fourth day after burial, an anatomist and his helpers dug up the body to dissect it in their laboratory. But after an incision had been made in the breast, the Englishman gave a cry and grasped the professor's arm! He had survived burial alive for four days and wrote a lengthy, sentimental account of his sufferings in the tomb, containing exclamations like "Each stroke of the hammer made my whole frame tremble. Oh, if I could only have cried out, or even given a sigh! But no!" The literary quality of the writing somehow suggests that this story was not meant to be quite *factual* and that the case report was intended as a horror story in a popular magazine.[16]

In the 1840s, the number of French newspaper reports of people being buried alive was greater than ever. It has been recorded that several French newspapers, including the influential *Le Siècle, Moniteur, Constitutionnel*, and *La Patrie*, were actively supporting the anti-premature-burial movement and that they were publishing every instance of alleged burial before death they could lay their hands on, without investigating its veracity (or lack of it).[17] In 1844, King Louis Philippe's government set up a commission to investigate the truth or falsehood of these rumors and to plan funerary reform if it turned out to be true that French people were frequently buried alive. Dr. Jules-Antoine Josat was appointed to travel to Germany to study the waiting mortuaries there. Dr. Josat was a thorough, if not particularly brilliant, medical scientist, and he took his assignment very seriously indeed. He made a lengthy tour of the waiting mortuaries in Germany. In 1854, after ten years of study, he put the fruits of his researches before the Academy of Sciences in Paris, hoping to receive one of the prizes.[18] Dr. Josat disagreed with Bouchut's proposal of the heartbeat as a certain sign of death; he claimed that colleagues of his had treated two cases of apparent death (an infant and a woman who had taken poison), where there was no audible heartbeat. In light of these two cases, the usefulness of Bouchut's treatise could indeed be questioned; maybe the glowing praise the academy had bestowed on him would prove just as *premature* as many a burial performed according to Bouchut's guidelines.

Dr. Josat was aware that many diseases, such as syncope, apoplexy, cerebral commotion, drowning, hysteria, and narcotic poisoning, could result in a state of apparent death. He condemned all traditional signs of death as uncertain: putrefaction had to be awaited. He had to admit that it reduced the impact of the many newspaper rumors about premature burials that out of more than 46,000 corpses taken into the German waiting mortuaries, not a single one had revived. But in France, where 800,000 people died annually, there could not be too many precautions to protect the helpless apparently dead from the horrible fate that awaited them. Our old friend M. Le Guern had given Josat a list of forty-six French cases of premature burial that had occurred in twelve years, and although he elsewhere demonstrated a healthy mistrust of the arguments of the anti-premature-burial activists, Dr. Josat was for some reason greatly impressed with this figure. He proposed that in addition to obligatory certification of death by a doctor in all of France, and a twenty-four-hour delay between death and burial, waiting mortuaries like that in Frankfurt were needed all over the country. He ended his book by claiming that all his many years of hard work and numerous sacrifices would have been worthwhile if he could remove the formidable scourge known as premature burial from the people of France. It is apparent that he was also very keen to receive a prize from the

Academy of Sciences, however, for the front page of his book proudly carries the inscription "Couronné par l'Institut." But the optimistic doctor had probably added this inscription before the prize committee had actually met; he received only an honorable mention and an *encouragement* of one thousand francs, and the committee took great exception to his expansive and unrealistic building program for waiting mortuaries.[19] Even an English reviewer was impressed with the thoroughness of Dr. Josat's work, although he deplored that "Sanitary police, at least in England, is indifferent about the risk of a few burials alive, and thinks it superfluous to prevent their occurrence."[20]

In 1866, the French anti-premature-burial campaigners presented another appeal for burial reform before the Senate of the Second Empire.[21] It had been devised by Mme Du Fay and M. de Carnot and was supported by several influential politicians. They demanded that the time between declaration of death and burial should be extended from twenty-four to forty-eight hours and that waiting mortuaries, with equipment for electrical resuscitation of corpses, should be built in every town. Two distinguished senators, M. Tourangin and the vicomte de Barral, took part in the debate, recounting some horrific newspaper stories about gnawed fingers, bloody shrouds, and convulsively clenched fists full of human hair to try to curdle the senators' blood. A distinguished prelate, Cardinal Archbishop Donnet, then stood up to speak. He claimed that he had saved the lives of two people presumed dead. He first recounted the tale of a woman who had recovered from her trance when her coffin was carried to the cemetery to be buried. This was mild stuff compared with the *grand guignol* scenes conjured up by the vicomte, but what was to come was more impressive. The cardinal spoke about a young priest who had collapsed in the pulpit while delivering a sermon. Taken out of the church and carried home, the priest could neither see nor move a muscle, but he could still hear, and what he heard was not apt to reassure him. After the doctor came and pronounced him dead, he could hear the funeral bell tolling. The venerable bishop, in whose cathedral the young priest had been preaching, came to recite the De Profundis. His body was measured for a coffin, and the next day his friends came around to see him for the last time before his coffin was carried away. But when the young priest heard the voice of an old friend he had known from infancy, he made a superhuman effort to speak and succeeded in making his resuscitation known to the bystanders. He recovered so quickly that a day later he again stood in the pulpit.

Before some incredulous senator could intercede with a demand that this case be formally proven, or that he was not impressed by this histrionic priest's swooning away, the dignified cardinal continued, "The young priest, gentlemen, is the same man who is now speaking before you!" In the stunned silence that ensued, he urged them to learn by the example he had put before them that diagnosis of death by a doctor was still fallible and that burial reform was needed to avert the most fearful and irreparable of misfortunes.[22] The Senate rejected the proposition, however; another appeal in 1869, in which Cardinal Donnet again took part, had as little effect, although the debate was lengthy and the newspaper coverage extensive. Regarding the 1866 and 1869 appeals, none less than Charles Dickens commented that the French arguments were founded more on sentiments than on facts.[23] There had been no case of apparent death in the German waiting mortuaries, and people were becoming unwilling to deposit their dead there.

Although new proposals for burial reform were submitted under the Second Republic, in 1879, 1880, and 1883, they lacked support and were easily brushed aside.[24] By this time, the old anti-premature-burial stalwarts, like Hyacinthe Le Guern, Hortense Du Fay, and Léonce Lenormand, were either dead or getting old, and the movement had difficulties recruiting replacements who possessed the same monomaniacal energy and unflagging interest in preventing premature interments. Furthermore, very few medical men were by this time on the side of the anti-premature-burial movement.

Eugène Bouchut worked for thirty years to undermine the public fear of premature burial. Already in the 1849 edition of his book, he had disproved some of the old idle rumors peddled in the French activist pamphlets: for example, he discovered that the Careless Anatomist stories about Vesalius, Philippe Peu, and the doctor of the abbé Prévost were without factual foundation. He also elegantly disposed of four newspaper tales from one of M. Le Guern's "lugubrious volumes"; he had written to the mayors or town clerks in the cities involved to ask for verification, and the stories all turned out to be completely false. For instance, an account in the influential *La Presse* of December 12, 1846, had confidently stated that a man had knocked inside his coffin in the town of Cluny, Saône-et-Loire; when his coffin was opened, he leapt up and ran away, jumping over the cemetery wall. The mayor of Cluny replied that the factual basis for this story was that the named man had died of pleurisy and that the doctor in charge had prudently waited eight hours before declaring him dead. This had started the rumor, which grew steadily while working its way toward the big national newspapers.[25] On November 29, 1867, *Le Figaro* contained another dramatic resurrection story, which again proved utterly untrue; the local authorities wrote to Dr. Bouchut that the individual in question had died more than a year earlier, without any untoward incident.[26] In the 1883 edition of his book, Bouchut doubted the existence of contemporary premature burial altogether—it was just a newspaper myth—and he recommended that all doctors investigate and disprove alleged cases. Other French skeptics, including the influential professor Paul Brouardel, followed his example. After twenty years of investigation, Brouardel could conclude that not a single newspaper report of live burial could be verified from the relevant local authorities; in many instances, he suspected that unscrupulous journalists had themselves invented their stories or fetched them from old newspapers.[27]

In 1862, the Dutch professor Alexander van Hasselt published the first German-language book that flatly contradicted Hufeland's doctrine of the *Scheintod*: premature burials, he wrote, were not common, and the signs of death, as described by Dr. Bouchut and other proponents of modern medicine, were perfectly reliable.[28] He did not go as far as the German Georg Varrentrapp and the Frenchman Gaultier de

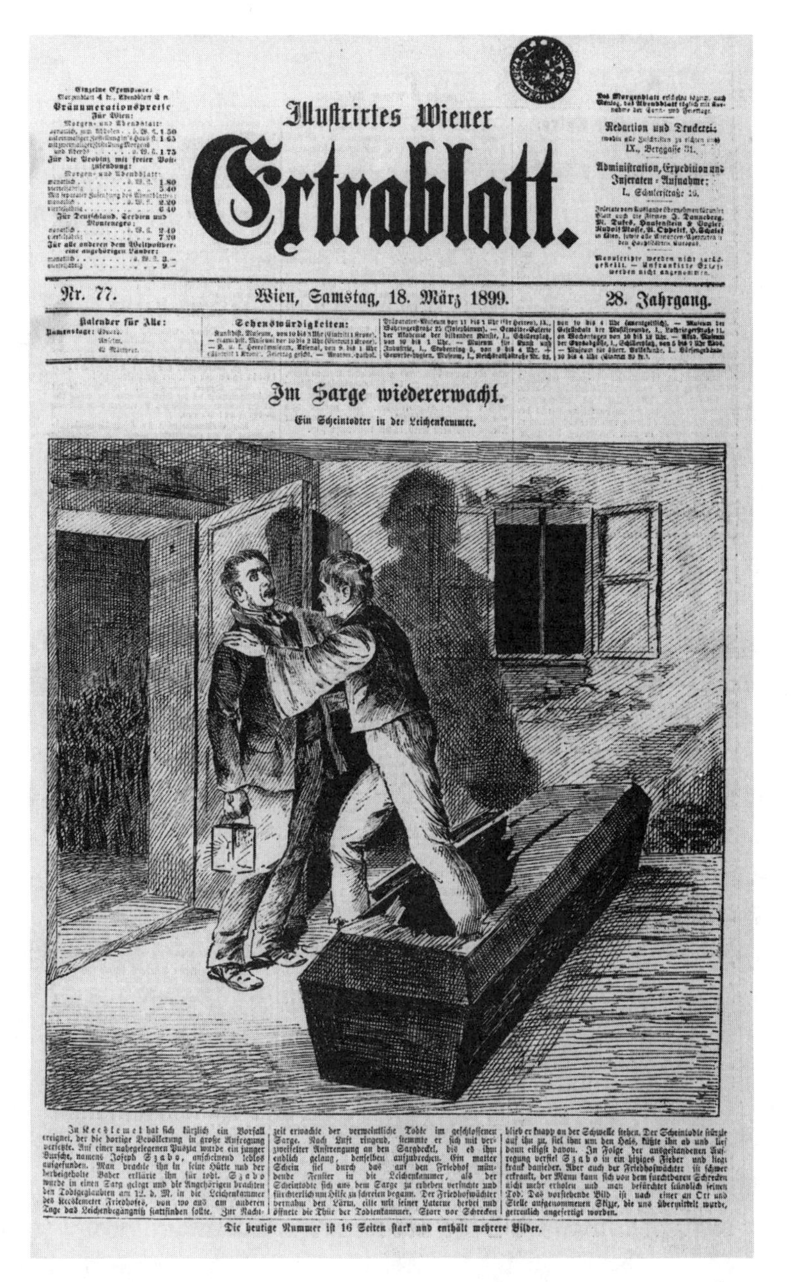

Illustrirtes Wiener

Extrablatt.

Nr. 77. Wien, Samstag, 18. März 1899. 28. Jahrgang.

Im Sarge wiedererwacht.

Ein Scheintodter in der Leichenkammer.

In Kecskemet hat sich kürzlich ein Vorfall ereignet, der die dortige Bevölkerung in große Aufregung versetzte. Auf einer nahegelegenen Puszta wurde ein junger Bursche, namens Joseph Szabo, anscheinend leblos aufgefunden. Man brachte ihn in seine Hütte und der herbeigeholte Bader erklärte ihn für todt. Szabo wurde in einen Sarg gelegt und die Angehörigen brachten den Todtgeglaubten am 12. d. M. in die Leichenkammer des Kecskemeter Friedhofes, von wo aus am anderen Tage das Leichenbegängniß stattfinden sollte. Zur Nachtzeit erwachte der vermeintliche Todte im geschlossenen Sarge. Nach Luft ringend, stemmte er sich mit verzweifelter Anstrengung an den Sargdeckel, bis es ihm endlich gelang, denselben aufzubrechen. Ein matter Schein fiel durch das auf den Friedhof mündende Fenster in die Leichenkammer, als der Scheintodte sich aus dem Sarge zu erheben versuchte und fürchterlich um Hilfe zu schreien begann. Der Friedhofswächter vernahm den Lärm, eilte mit seiner Laterne herbei und öffnete die Thür der Todtenkammer. Starr vor Schrecken blieb er knapp an der Schwelle stehen. Der Scheintodte stürzte auf ihn zu, fiel ihm um den Hals, küßte ihn ab und lief dann eiligst davon. In Folge der ausgestandenen Aufregung verfiel Szabo in ein hitziges Fieber und liegt krank danieder. Aber auch der Friedhofswächter ist schwer erkrankt, der Mann kann sich von dem furchtbaren Schrecken nicht mehr erholen und man befürchtet stündlich seinen Tod. Das vorstehende Bild ist nach einer an Ort und Stelle aufgenommenen Skizze, die uns übermittelt wurde, getreulich angefertigt worden.

Die heutige Nummer ist 16 Seiten stark und enthält mehrere Bilder.

A man has awakened from his death trance in the *Leichenhaus* of Kecskemét and stands up in his coffin to embrace the startled watchman. A drawing alleged to be based on a real event, from the *Illustriertes Wiener Extrablatt* of March 18, 1899. *Reproduced by permission of the Österreichische Nationalbibliothek, Vienna.*

Claubry and deny the occurrence of premature burials altogether, but he was confident that they were very rare. While active in Utrecht for twenty-four years, he had not come across a single case; nor had Dr. Josat, after a similarly lengthy period at the Paris morgue, encountered any instance of live burial. Professor van Hasselt was also aware of the role of the French newspaper propagandists in the opposition to premature burial and could demonstrate that their untrue stories spread from newspaper to newspaper, even reaching respectable medical publications. In 1851, he had been interested to see a case of premature burial from his hometown of Utrecht quoted in a French medical pamphlet. A prematurely buried cholera victim had made his resuscitation known too late and was found soaked in blood in the coffin; he had eaten three of his fingers in his agony. The source was the respectable medical journal *Gazette des Hôpitaux*, but this periodical had reproduced it from a Dutch newspaper, the *Kamper Courant*. Inquiring into the circumstances at the relevant hospital and cemetery, van Hasselt found the story to be completely untrue. Similarly, the German professor Göppert had investigated a series of newspaper accounts of alleged premature burials in Rheinhessen; all were complete inventions except one, which had only concerned a corpse that had been warm longer than usual in the local mortuary.[29] In 1867, a Danish doctor could demonstrate that four out of five local tales of apparent death and premature burial reported in the *Dagbladet* newspaper were total fabrications.[30]

A series of French articles published in 1880 gives an interesting insight into how these popular tales of premature burial originated, and how they spread through folklore and the newspaper media.[31] A certain Dr. P. Keraval wrote that he had once been called in to examine the contents of three coffins in a morgue, after a workman said he had heard knocking. All three corpses turned out to be putrid. But it did not take many days before the entire town knew that one of the municipal councilors, whose remains had been in one of the coffins, had been buried alive. Nor was this tale repudiated by the local newspapers. Dr. Keraval wrote a denial, but was unable to convince people of the truth.

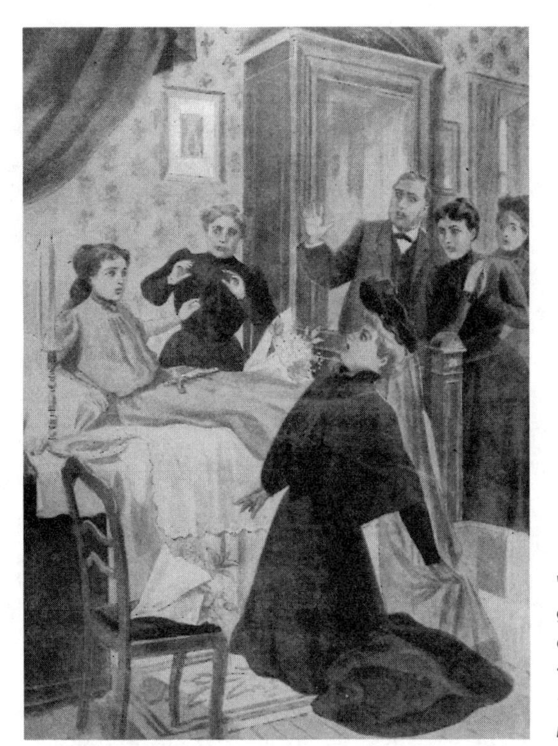

"La Morte Vivante": a young girl awakens after being declared dead. An illustration from *Le Petit Journal*, 1902. *Copyright J.-L. Charmet Photographic Library, Paris.*

He cynically had to conclude that his patients, who were busy gossiping that one of the local worthies had been prematurely buried because of his neglect, trusted the newspapers more than their doctors.

Dr. A. Job, of Lunéville, had had an even more dismal experience. A patient of his had died from tuberculosis, at the age of twenty-six. In the evening of the next day, he was called by the mayor's son. White-faced and distraught, this gentleman told the doctor that while digging another grave, the verger had heard a sinister knocking sound from the man's recently buried coffin. The coffin was of course exhumed and the doctor sent for at once. After a long, cold ride through heavy snow, Dr. Job arrived at the churchyard, where twelve people were solemnly standing around the man's grave. The local schoolmaster, who also claimed to have heard the knocking, was telling everyone that the man had been buried alive. But when the coffin was opened, the corpse was ice cold and still, and there were no signs whatsoever of an awakening in the tomb. Dr. Job presumed that the knocking sound was the result of the swelling of the coffin boards. The verger stubbornly shook his head when told this, and before long all of Lunéville knew that their doctor had buried one of his patients alive.

A certain Dr. E. Decaisne had also fallen afoul of the unreasonable obsession with premature burial that reigned in France at this time. When an elderly lady, who lived with her sister in a large château, died, Dr. Decaisne was called in to examine the body. Her sister refusing to believe that she was really dead, urged that the lifeless body be "stimulated" in some way. Although the doctor found this distasteful, he took a red-hot poker and burned the corpse over the face. The old woman then flew at him like a maniac, screaming that he had disfigured her sister. Only after Dr. Decaisne had sworn a solemn oath, on the Bible, that the woman was really dead, could her sister be persuaded to let go of the corpse and to allow it to be shrouded and coffined. Several years later, a workman came to Dr. Decaisne and urged him to cut his recently dead wife's head off before she was buried. When the doctor wished to know the reason for this strange request, the workman said that many local people knew that one of the doctor's other patients—the old lady at the château—had been buried alive.

IT WAS TOLD that the daughter of Henry Laurens, the first president of the American Continental Congress during the Revolutionary War, was declared dead of smallpox when quite young, was shrouded and coffined, but awoke from her trance, recovered, and lived to a mature age. This narrow escape made such an impression on her father that he lived in fear of a premature interment and directed in his will that his body be cremated. Another American worthy, General Robert E. Lee, had similar concerns, also, it was claimed, brought on by a tragic occurrence in his close family. His mother was prone to seizures, and was declared dead by a physician after one of her fits. But when the sexton was filling in the grave, Mrs. Lee recovered, knocked on the coffin lid, and gave a loud cry. She was rescued in time.[32] Still, the European fear of premature burial took quite a while to reach the United States. The books of Taylor and Snart were not published in America, nor were there any American translations of the numerous French and German pamphlets on the subject. Early nineteenth-century American medical science was much influenced by the British tradition and thus escaped Hufeland's doctrine on the *Scheintod*. Although the 1820s and 1830s American medical articles on the diagnosis of death agreed that putrefaction was the most certain sign of death, the other signs were not considered worthless, if used with caution.[33] The diffusion of the fear of premature burial into American society thus occurred not through medical science but through popular books and magazines. Edgar Allan Poe had no difficulties in procuring material for his study on the subject: *Blackwood's Magazine* delighted in publishing stories about premature burials, as did the appositely named *Casket* and several other American newspapers and magazines.[34] Both the Lady with the Ring and the Two Young Lovers appear in several guises, along with the usual bugaboo tales of gnawed fingers, bloody shrouds, and horribly contorted bodies found inside coffins. In 1847, the first appeal to build waiting mortuaries on American soil appeared, after an article reproduced in the *New York Observer* and several other papers had praised the German mortuaries for their sterling work for the prevention of premature burials.[35] The appeal was unsuccessful, although it ended with these stirring words:

> To awake from what may seem the sweet sleep of returning health, and find one's self not only dressed in the habilments of the grave, but inclosed within its remorseless grasp—to feel the sickening pang which the first realization of this horrible fact produces—to struggle when no hope cheers the effort—to hear perhaps the hum of life as it rolls over our heads, and from which we are cut off forever—to call out when no ear can be reached—and finally give up and wish to die, but cannot be released—the idea is replete with horror, and the risk should never be permitted.

In 1871, a more serious appeal was launched by Dr. Alexander Wilder, who read an address on the perils of premature burial before the New York State Assembly.[36] He told some of the usual horror stories of corpses found in unnatural positions, and demanded a change in the legislation, so that no person could be buried without being certified dead by a competent physician. Wilder was a proponent of eclectic medicine, which encompassed all kinds of dubious and murky theories on herbalism, self-healing, and homeopathy. He rejected the bacterial origin of disease, preferring the time-honored dogma that disease was due to overexposure to cold, sexual excesses, unwholesome postures in bed, or mental over-excitement. Wilder was an antivaccinationist, thus denying the merits of one of the greatest medical breakthroughs available at his time. He was also a spiritualist, and many of his friends in this movement shared his fear of being prematurely buried. Since they believed that the spirit could leave the body at will to go and gad about on its own, they found it precarious, to say the least, that the spirit might have to return to a body that had been declared dead from lack of an occupant, shrouded and buried under six feet of earth. From the 1870s on, the anti-premature-burial movement in Britain and the United States was to recruit a fair proportion of its members from the spiritualists.

One of Alexander Wilder's followers in the eclectic medical movement was Dr. Moore Russell Fletcher, of Boston. In 1883, he published a book entitled *Our Home Doctor*, which combined a guide to family health with his pet theories on herbalism and homeopathy.[37] In between sections on delirium tremens, tapeworms, and ingrown toenails, we find a discussion on whether alcoholic liquors, sexual excitement, or superabundance of bile is the most frequent cause of disease, as well as the information that the Africans missed the train of evolution and thus remain the same as they were thousands of years ago. After 332 pages of this nonsense, we reach the pièce de résistance: the book also contains a pamphlet entitled *One Thousand People Buried Alive by Their Best Friends*. Human beings could enter a state of hibernation just like snakes or toads, Dr. Fletcher wrote, and this made them liable to enter a state of trance deep enough to be mistaken for death. In fact, his own aunt Nancy had been buried alive. The old European legends of premature burial are not spicy enough for Dr. Fletcher, who actually improved on them. We learn that "Andrew Vesale" deliberately murdered his own patient, by stabbing him in his still-beating heart, to cover up his fatal error of declaring him dead; the wicked anatomist then fled to Asia. The Lady with the Ring story is placed at Toulouse in 1866 and ends fatally for the robber. Dr. Fletcher later edifies the reader with a few American case reports culled from the newspaper columns. In May 1869, a young lady in New Orleans was certified dead by her own uncle, a physician, after she suddenly fell ill. She was buried in a vault that same afternoon, but the sexton's son heard the most dreadful sounds from the vault and called his father. When her parents arrived at the cemetery, no sounds were coming from the vault. After it was opened, it was seen that the girl had struggled fearfully: her hands were clenched and bore marks of teeth, her hair was torn from her head, and her face was terribly distorted. "The horror of the parents, and of the physician and uncle, cannot be described." Equally pleasant was the tale of Mrs. Crane, the wife of a bookkeeper, who died, or so it was thought, in 1868. Her mother grieved her death bitterly, and the sorrow was believed to have affected her reason. One day, she came rushing into her son-in-law's room screaming, "You have buried my daughter alive! What shall I do!?" To allay her fears, the tomb was opened, but the sight that awaited the wretched mother was not a prepossessing one: there was blood everywhere, the glass pane in the coffin lid was broken, the corpse's fingers were bitten and cut by glass, and the hair almost entirely torn out of the head. After this shock, Mrs. Crane's mother herself fell dangerously ill, and her life was feared for.

Much of the American anti-premature-burial agitation was centered in New Jersey and Massachusetts: Alexander Wilder was active in Newark and Moore Russell Fletcher in Boston. The editors of a magazine entitled *Our Dumb Animals*, published in the latter city, were concerned both about for poor dogs tortured by cruel vivisectors and about apparently dead people at risk of being buried alive. The spiritualist, antivivisectionist, antivaccinationist, and homeopathic communities in this part of America were closely linked, and many of their members also agitated against premature burials. In 1889, the singular death of one of their leading members led to immense outrage throughout these communities. Washington Irving Bishop, the famous spiritualist and mind reader, had collapsed after one of his shows in New York and been pronounced dead by a doctor the following night. He was suffering from some kind of cataleptic condition, and it was said that he always carried a note saying that if he was considered dead, his body should be kept in a safe, warm place and his own doctor be sent for. Under no conditions should it be autopsied. When Bishop's wife went to the Hawkes Funeral Parlor on Sixth Avenue to see his lifeless body, she wanted the attendant to comb his hair. But as the man did so, he dropped the comb, which fell into the empty cranium. Bishop had been autopsied, and his brain was missing. Mrs. Bishop started screaming and accused the doctors in charge of having murdered her husband to get his brain, which was thought to be particularly developed as a result of his mind-reading talents. Bishop's mother, the eccentric spiritualist Eleanor Fletcher Bishop, also charged the doctors with murder by premature autopsy. At the inquest, a verdict of not guilty was returned against the three doctors involved, although one of them was reprimanded for the undue haste of the autopsy. Bishop's brain had actually been found, inside the corpse's thoracic cavity. The outraged Eleanor Fletcher Bishop then published a pamphlet entitled *Human Vivisection of Sir Washington Irving Bishop*, illustrated with a portrait of herself embracing her son's lifeless body in its casket.[38] She boldly accused the doctors of having murdered her son: "Dr. Frank Ferguson (the Jack-the-Ripper of America) *ripped my angel boy open, and took out his heart, and sawed his beautiful fair brow and head into two pieces, took out his wonderful brain*, and then the *butchery* of my only child *was accomplished. . . .*" The Bishop case, often mentioned in American anti-premature-burial agitation, caused much sensationalist press commentary. Alexander Wilder and his fellow activists all agreed that Bishop had been autopsied while still alive.[39]

In 1895, the Occult Publishing Company in Boston brought out a book entitled *Buried Alive,* by Dr. Franz Hartmann, M.D.[40] This deplorable volume contained a greater proportion of horrid stories than any previous work in the English language, outclassing even John Snart's *Thesaurus of Horror*. Hartmann was no ordinary doctor: he was a spiritualist and friend of Mme Blavatsky, a Freemason, and a Rosicrucian, cofounder of the Ordo Templi Ordinis magical society, and inventor of a bogus inhalation therapy for tuberculosis. A native of Austria, he had emigrated to the United States in the 1860s and served as a medical officer in the American Civil War. Later, he traveled around in the southern states, practicing as an itinerant doctor (some would say quack) and for a time serving as coroner in Georgetown, Colorado. Many of Franz Hartmann's sanguinary case reports were culled from the pages of the similarly spirited work of Léonce Lenormand. Hartmann published three versions of the Lady with the Ring, from Salzburg, Kronstadt, and Beaujolais, and three versions of the Two Young Lovers. The Austrian newspapers supplied further material. In a small town in the province of Styria, a young pregnant woman died and, after a wait of the customary three days, was buried in the local churchyard. Some days after the burial, it was rumored that her husband had poisoned her, and the grave was opened by order of the police. There was evidence of a terrible struggle, and the woman had given birth to a child in her coffin. The physician who had signed her certificate of death was sentenced to a few weeks of imprisonment, in the hope that a period of incarceration would improve his diagnostic ability. Hartmann considered all signs of death as highly fallible, since a spirit could leave the body and reenter it at will. Even putrefaction was uncertain, because he could reproduce some dubious German instances of "corpses" that emitted an insupportable odor, but still came to life again. He had himself observed "a negro in Texas" who "emitted a terrible odor like a corpse in an advanced state of putrefaction." A correspondent of his, Mr. H. R. Phillips, of 51 East 50th Street in New York, wrote that one of his friends, a sea captain, came home to find that his wife had died in his absence. He had the body exhumed and, from the position of the body, grew convinced that she had been buried alive. As a result of this horrid experience, he became insane and was confined in an asylum; the wife's sister, the clairvoyant Millie Fancher, of Brooklyn, was affected even worse: she took to her bed and never rose from it again, living entirely on liquids for many years.

Franz Hartmann's book was later reprinted in London, but with little success: not a single copy seems to be kept in any British library. The London Society for the Prevention of Premature Burial bought a number of copies, but buyers were so scarce that they finally had to give the books away to newly recruited members. The reviewer of Hartmann's book in the *British Medical Journal* was scathing when he quoted one of the more fanciful case reports (actually culled from the pages of Léonce Lenormand): "What is to be thought of the credibility of an author who gravely narrates the case of an Englishman who died of typhus fever in 1831, who was buried four days later, and after another four days of burial in a coffin in a grave, was exhumed and found alive, and who stated that he had been conscious all the time, and that his lungs had been paralyzed and used no air, and that his heart did not beat." The reviewer concluded bluntly, "A worse farrago of nonsense than that contained in this pamphlet it has seldom been our lot to come across."[41]

IX

The Final Struggle

> I dare not breathe the sequel loud,
> 'T would horrify lone folk
> Of how they draped me in my shroud
> 'Mid coarse and brutal joke.
> And how upon my coffin lid
> I heard the hammer's stroke!

The verses quoted above and later in the chapter are from *Living with the Dead*, a truly dreadful, thirty-two-stanza poem by a certain Mark Melford, published in the magazine of the organized anti-premature-burial activists in England.[1] This organization was not the first of its kind. It is recorded that a society was formed in Weimar, just before the first waiting mortuary was built, with the express purpose of safeguarding its members from premature interment. Similar bodies existed in

Germany throughout the nineteenth century: they employed doctors to watch the bodies of deceased individuals in the morgue until the onset of putrefaction, or sometimes also to open an artery after death had been confirmed. One was founded by Colonel von Falkenhausen in Breslau in 1860, and another was later active in Berlin, campaigning for cremation as the only safeguard against premature burial. In France, a body of concerned parties formed after Cardinal Donnet's appeal in 1866, and in America the agitation of Alexander Wilder had a similar result. None of these organizations made much of an impact, however, and little is known about their activities. The London Society for the Prevention of Premature Burial was formed in 1896 by Mr. Arthur Lovell, who was a somewhat shady character: a spiritualist, quack, and author of books on self-healing and other forms of medical humbug. Lovell and the cofounders of the society set out its three aims: the scientific study of apparent death and catalepsy, popular agitation to make people aware of the risk of being prematurely buried, and a service to members that doctors employed by the society would make sure that they were never buried alive.

The society's leading light was a somewhat more respectable figure: the traveler and controversialist William Tebb.[2] Under his leadership, it expanded over the Atlantic and also gained members from continental Europe. Born in Manchester in 1830, Tebb became a political radical at an early age and spent his youth agitating against the Corn Laws and fiscal protectionism. He later moved to Massachusetts, where he became known as a fierce abolitionist and a personal friend of the antislavery apostle William Lloyd Garrison. After the American Civil War, Tebb turned his attention to what he perceived as the evils of vaccination and wrote two books attacking these practices. In 1879, he and Alexander Wilder founded the Anti-Vaccination Society of America. The struggle of these antivaccinationists to stop the use of the first real advance in preventive medicine was supported by a motley crew of quacks, homeopaths, and patent-medicine manufacturers. They claimed that vaccination was a bluff, and that millions of people were inoculated with a loathsome pestilence and were doomed to carry to

Two anti-premature-burial propagandists: Mr. William Tebb (*left*) and the sinister-looking Dr. Franz Hartmann. *From W. Tebb and E. P. Vollum,* Premature Burial and How It May Be Prevented *(London, 1905), and the* Burial Reformer *magazine of July–September 1905, respectively.*

their graves bodies wasted with cancer and consumption. One of Tebb's antivaccinationist colleagues wrote that vaccination introduced into the bloodstream "a bioplasm, death-laden, carrying with it all the vices, passions and diseases of the cow." He put vaccination on a par with alcohol, tobacco, lust, and sensual love as the most destructive elements in society.[3] The fact that Tebb had no medical education whatsoever did not prevent him from arguing his case with vigor; inevitably, the medical community regarded him as a dangerous reactionary and faddist. He arranged public demonstrations against compulsory vaccination in various British towns, walking under banners with captions like "It is not smallpox you are stamping out, but human creatures" and "Better a Felon's Cell than a Poisoned Babe." Tebb then became interested in leprosy and traveled widely to study this disease. In 1893, he published a book that presented a case for the inoculability of leprosy and its communication by means of vaccination.[4] Before his unrequited love for medical science began, Tebb had become a fellow of the Geological Society, and the letters F.G.S. were flaunted on the title pages of his books as if they signified a medical degree. In addition, the Royal Academy of Medical Sciences of Palermo had shown the bad judgment to make this sad travesty of a medical scientist a corresponding member, and Tebb proudly displayed this dubious distinction in the same manner.

When living in India in the early 1890s, Tebb met a distinctly odd sanitary physician named Roger S. Chew, who told him a long and woeful tale of how his entire life had been haunted by the ghost of premature interment.[5] When he was a schoolboy at Bishop's High School in Poonah, his sister died of convulsions. Roger was so distraught that he entirely lost the power of speech, refused to eat, and gradually fell into a decline. When he attended her funeral, he threw himself headlong into her grave with a dismal howl and was dragged out unconscious. He did not regain consciousness until several days later, but then had a relapse and was declared dead on January 18, 1874. When he was laid out for burial in a coffin, his other sister declared that she had seen his lips move, and a doctor was called. After a large abscess of the neck had been drained, young Roger's life was saved. Following a lengthy convalescence, Roger Chew decided to study medicine and in 1877 became an assistant surgeon in the Indian army. While traveling on a bullock train with three other surgeons, one of his fellow medical practitioners "showed strong symptoms of cholera" and was declared dead a few days later. His coffin was put in a hospital dispensary, but the day after, the young doctors were startled to find that their friend had come back to life. Exactly the same thing later happened to two soldiers in the East Norfolk regiment; had there not been a shortage of wood for their coffins, Dr. Chew asserted, they would have been buried alive. Chew was often employed in the exhumation of old graveyards, and it happened more than once that he found corpses in odd positions in the old coffins and vaults; his diagnosis was of course that they had been buried alive and had tried to dig or break their way out. But worse, much worse, was to come!

A great friend of Roger Chew's named Frank Lascelles fell headlong forward onto his plate during dinner, "in the middle of a burst of hearty laughter," and was declared dead. Six months later, Chew obtained permission to remove his friend's remains to another cemetery. On removing the lid from the coffin, he was horrified to note that the position of the mummified corpse indicated that a fearful struggle had taken place underground. Another, more experienced doctor tried to explain matters by suggesting that the jolting of the coffin on the way to the cemetery might have overturned the body, but Chew was unconvinced. In 1881, one of his relations, Mr. J. A. A. Chew, was buried in the family vault in Calcutta. Roger could not resist the opportunity to have a look around in the vault. He saw that the lid of the coffin of one of his female relations, the seventeen-year-old Mary Norah Best, had been torn off and that the skeleton was situated half in, half outside the coffin. Roger Chew believed that the family surgeon had twice tried to murder young Mary's mother, "who fled from India to England after the second attempt on her life, but, unfortunately, left the girl behind." The wicked surgeon had poisoned the girl and then certified her dead, but inside the vault "she recovered consciousness, fought for life, forced her coffin open, and sitting up in the pitchy darkness of the vault went mad with fright, tore her clothes off, tried to throttle herself, and banged her head against the masonry shelf until she fell forward senseless and dead." In 1894, Roger Chew exhumed the body of an eleven-year-old girl, to whom chloroform had been administered under somewhat peculiar circumstances shortly before death. His diagnosis was of course that she had been buried alive while in a cataleptic condition caused by the chloroform: "What a mournful vista little Sarola's case opens up, and who can say how many hundreds have been similarly disposed of!"

The ears of all it would appall
Were I these things to tell,
Bold men would quake did I relate
My ghastly funeral!
These hideous acts, these loathsome facts
Should be proclaimed in Hell!

IT IS NOT known whether William Tebb was alarmed by the risk of premature burial before his meeting with Dr. Chew, but it appears that these gloomy revelations from the sanitary physician prompted him to study this subject in detail. Tebb later met the American colonel Edward Perry Vollum, M.D., a former inspector of the U.S. Army Medical Corps, and they found a mutual interest in the premature burial controversy. As we know, Tebb was something of a busybody who dabbled in all kinds of medical faddism, but Vollum was a respectable military surgeon who had served with distinction in the American Civil War and later held important medical posts in the U.S. Army for a third of a century. Once, as a child, Vollum had been laid out for dead after drowning, but he awoke on a board in the mortuary chamber, surrounded by corpses on all sides. Not surprisingly, this ghastly experience affected him very much, and he decided to cooperate with Tebb in writing a book to save others from being buried alive. Tebb, an experienced author, did most of the writing; their book *Premature Burial and How It May Be Prevented* was published in 1896.[6]

The first edition of this book had more than 350 pages and must have been quite difficult to shrug off at the time. Even today, it is quite an impressive tome: clearly written, mostly accurate in details and references, and passionate in its arguments in favor of burial reform and construction of mortuaries. With characteristic energy, William Tebb had scanned newspapers for recent sensational cases of premature burial, without making any attempt to verify them. He also had a talent for extracting those parts of medical articles that aided his crusade against premature burial, as some unwary medical writers in the *Lancet* and the *British Medical Journal*, who had disapprovingly quoted Continental alarmist sentiments, would find out to their chagrin. Tebb presented the quotations as the opinions of the writers and thus recruited some traditionalist medical men for his cause. Not knowing that the German waiting mortuaries were actually in the process of removing their signaling apparatus, he contrasted these impressive,

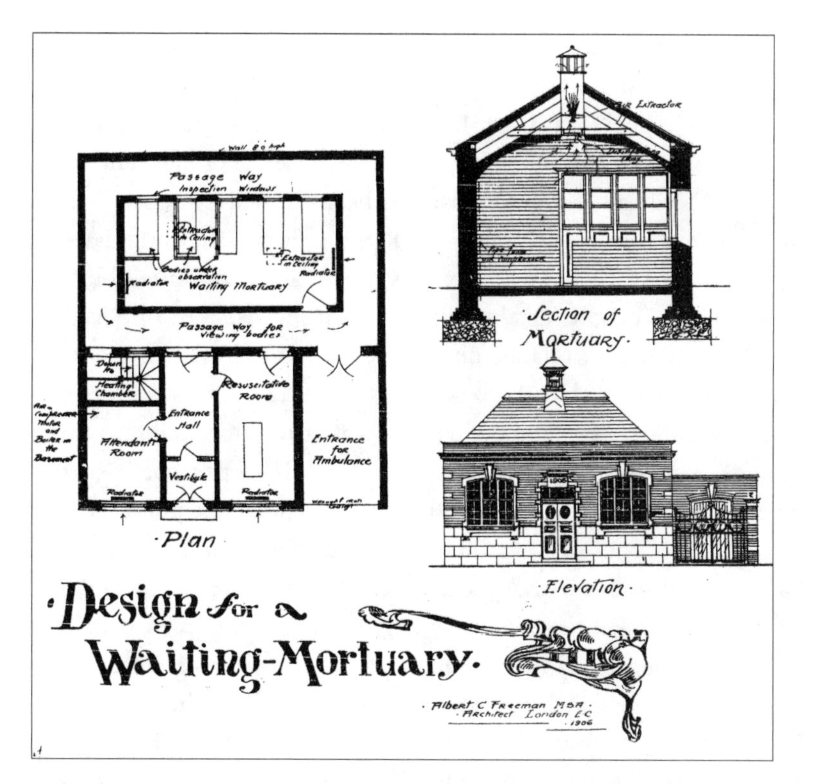

A plan for a grand waiting mortuary, by Mr. Albert C. Freeman, M.S.A., from the *Burial Reformer* magazine of April–June 1906.

humanitarian establishments with the plain, gloomy mortuaries of London, which lacked any form of supervision and put the inhabitants of the metropolis in imminent danger of being buried alive.

The medical advances of the late nineteenth century had added some arguments in favor of the uncertainty of the signs of death, among them the discovery that cardiac massage could restart a heart that had stopped beating and that anesthesia could produce a death trance resembling the state described by Hufeland and others. But Tebb was largely ignorant of medical science, and Colonel Vollum was not up to date with modern physiology and scientific medicine. As a consequence, they were completely unable to appreciate the medical advances that in fact aided their cause. Their review of the French debate on the signs of death was both feeble and biased: they quoted the opinions of Josat and Icard with approval, but claimed that Bouchut's unfounded suggestion that the absence of an audible heartbeat was a certain sign of death had been refuted by reliable experts. Roger Chew was of the opinion that the vultures that circulated over the Towers of Silence, near Bombay, possessed some infallible means for telling the dead from the living, since they attacked only cadavers. Tebb preferred a less dangerous species of animal diagnostician, however: he suggested that the dog possessed extraordinary powers to recognize if some person laid out as dead was still living, and that the dead person's dog should always be brought to the deathbed, to detect whether a spark of life was still lingering. To prove his point, he quoted a tragic case where the verdict of the sagacious canine diagnostician had been left unheeded.[7] A postmaster in Moravia had died, and his little pet dog refused to leave the grave, but lay there howling piteously. Even if the dog was forcibly removed, it returned as soon as it could escape. This lasted for a week and became the talk of the village. A year later, when the church was extended, the postmaster's coffin was exhumed. The body was found in a dreadful state, leaving no doubt that the man had been buried alive; the physician who had signed the certificate of death went insane.

In fact, the only major difference between Tebb and Vollum's book and Bruhier's 150 years earlier was that the number of gruesome case reports had increased. Nor was Tebb less credulous than his eighteenth-century fellow propagandist. The tale of the Two Young Lovers is twice recounted, with differently named protagonists. The Lady with the Ring legend recurs in four of the well-known historical incarnations, as well as two modern ones: from New York in 1889 and from Ireland in 1895. Far from suspecting that these many variations on the same theme might indicate that these tales belonged in the realm of folklore, Tebb and Vollum authoritatively stated that "a study of the facts of premature burial shows that the rifling of tombs and coffins to obtain valuables has in other instances revealed similar tragic occurrences."[8] Another amusing story concerned an Indian missionary named Schwartz, who awoke from a deep death trance in his coffin when he heard his favorite hymn sung over him before the last rites were to be performed; he made his resuscitation known by joining in the song.

William Tebb, who must have had some private means of supporting himself in his campaign against medical wrongdoings, sent out several hundred review copies to newspapers and magazines all over the world.[9] He also thought of another, quite clever ruse to advertise his book. A similar number of copies were sent to kings, magnates, politicians, and opinion leaders all over the world. Since it was considered impolite not to reply to a letter at this time, and even less polite not to thank the donor of a book, quite a few of the great men wrote back. Tebb then carefully excised the parts of the letters that could be construed to imply support for his cause, and used these extracts in his advertisements and campaigns. The guardedness of their comments implies that some experienced statesmen, like Prime Minister William Gladstone and President Theodore Roosevelt, were wise to this trick, which may well have been practiced by various authors; others, like Jerome K. Jerome, Count Leo Tolstoy, Prince Krapotkin, and the lord chamberlain to the king of Denmark, were more exuberant—and found themselves roped in as supporters of the anti-premature-burial movement.[10]

Alive! Within the jaws of death,
No fate was ever worse!
No enemy invoked on me
So terrible a curse!
Conveyed still living to my grave!
Within a funeral hearse.

TEBB AND VOLLUM'S *Premature Burial and How It May Be Prevented* was widely and respectfully reviewed on both sides of the Atlantic. Just like Bruhier in 1742, William Tebb was clever enough to know when he had come upon a good thing, and he and his friends in the London Society for the Prevention of Premature Burial did their utmost to create newspaper publicity for their case. Even the respectable conservative magazine the *Spectator* inserted an unsigned editorial on this subject, beginning, "There is probably no horror more universal, more intense, more soul-subduing, than the horror of being buried alive." The writer claimed that if ten ordinary men or women were asked what their worst nightmare was, it would be that of awakening in the coffin after a premature burial. Tebb himself wrote a letter to congratulate the editor of the *Spectator* on bringing up the need for burial reform; he stated that his own interest in the subject stemmed from a distressing case of premature burial in his own family. One of his colleagues from the London Society for the Prevention of Premature Burial, Mr. James R. Williamson, also had a letter published, in which he quoted Franz Hartmann's alleged seven hundred cases on file and told a version of the Lady with the Ring story, from Ireland.[11]

Williamson later wrote a similar letter to *Scientific American* magazine, demanding burial reform both in America and in England.[12] This time, he was answered by an American practitioner, J. F. Baldwin of Columbus, Ohio.[13] Dr. Baldwin was well aware of the considerable quantity of unproven newspaper stories about premature burial. He had himself investigated several of them and found that they were without exception completely untrue. For example, the *Columbus Evening Dispatch* of March 12, 1890, carried a story entitled "Prepared for the Grave," about a four-year-old boy reviving in the coffin. The boy was named, and so was the village in which his family lived. Dr. Baldwin went there and found the lad healthy and well; his family told him that the youngster had never been ill, and wondered at the origin of this untrue report. In the Philadelphia *Medical News*, the editor of the *Spectator* was roundly criticized for trying to "create sensation for the dull season" by reproducing these false stories of people being buried alive.[14] A prominent American undertaker had personally investigated every newspaper report of premature burial for the past ten years.

Although these reports had been both numerous and highly colored, in every instance they were found to be devoid of factual foundation. The editor of the *New York Observer* made a similar inquiry, directed to leading Presbyterians and other respectable citizens in whose neighborhood premature burials, graphically reported by the sensationalist newspaper press, were said to have occurred. Even though taking pains to interview relatives and other alleged witnesses, he could not verify a single case.[15] The Boston magazine *Our Dumb Animals* focused on animals threatened by cruel vivisectors as well as on apparently dead people facing the horrors of a living inhumation, and its distasteful stories were eagerly regurgitated by several provincial newspapers. But again a doctor challenged this publication fueling an old superstition and showed that, of three sensational examples of premature burial recorded in *Our Dumb Animals*, two were complete fabrications and a third was lifted from an article that ran in the same magazine nine years previously. He deplored the insulting charge against the American medical profession that they could not tell a living person from a corpse, and rightly claimed that a cause that had to rely on invented stories, or on dishing up as current news material that had been in cold storage for nine years, must be in a bad way.[16]

In Britain, several medical men also took exception to what they perceived as an old fallacy regaining support. They were particularly dismayed that, at a time when Germany was closing down its waiting mortuaries, and the French opinion for burial reform had been calmed, there was an upsurge in interest and concern for this subject in both Britain and the United States, which could directly be traced to the publication of Franz Hartmann's *Buried Alive* and Tebb and Vollum's *Premature Burial and How It May Be Prevented*. In 1897, Dr. David Walsh, a dermatologist at the Western Skin Hospital in London, wrote a long article in rebuttal of Tebb and Vollum, which was later published as a pamphlet.[17] Like Dr. Baldwin on the other side of the Atlantic, he was heartily tired of the many horrific and untrue newspaper accounts of premature burials, and annoyed at the success of Tebb and Vollum's book. Like Antoine Louis in his attempt to rebut Bruhier's successful book 155 years earlier, Walsh was outraged by Tebb and Vollum's criticism of the medical profession and their stubborn denial to accept any sign of death except putrefaction. Again like Louis, he was not particularly successful. Dr. Walsh was no scholar of European mythology, and he could not see through the fictional and/or legendary nature of some of the older cases in Tebb and Vollum's book; nor was he an expert on the normal postmortem changes of a cadaver, and thus could not discuss the lurid newspaper stories from the perspective of a forensic scientist. In his onslaught against Tebb and Vollum's book, Dr. Walsh scored heavily only once, but his success belonged to the realm of buffoonery rather than that of forensic medicine. Tebb had incorporated an account of a certain "Professor" Fricker, a hypnotist who had performed in London in 1895. At the Royal Aquarium, Westminster, he had buried his hypnotized assistant alive under eight feet of earth, a shaft to the coffin being left open to permit respiration. The assistant, Frederick Howard, had toured Britain in a coffin, to be "awakened from the dead" at the beginning of every show. Since then, however, he had sued the "professor" for unpaid wages and denounced him as a fraud. In May 1897, after a farcical trial that was liberally quoted from by Dr. Walsh, Fricker was sentenced to pay the amount due to his erstwhile assistant.

David Walsh was of course perfectly right that *Premature Burial and How It May Be Prevented* was a far from scholarly work: its authors were hopelessly ignorant in physiology and forensic medicine, and the majority of their cases taken from "reputable writers from time immemorial" rested on very weak evidence indeed. Dr. Walsh's refusal to accept any of the reported cases of premature burial is unconvincing, however, and Tebb himself rightly observed that thirty-two live burials, and sixty-nine narrow escapes, in his book were quoted from medical sources.[18] Although Dr. Walsh's pamphlet was approved by the medical profession, there is no evidence that it even came close to alleviating the residual fears of premature burial that had been reawakened by Tebb and Vollum's book. In pamphlet form, it included a letter from the quack Arthur Lovell, secretary of the London Society for the Prevention of Premature Burial, who put forward the case of the rescue from the tomb of the countess of Mount Edgcumbe, one of the British Ladies with the Ring, as a scientifically proven example of premature burial. It was vouched for by Sir J. Tollemache Sinclair, Bart., formerly MP for Caithness, who had heard it from a grandson of the countess. In a scathing remark, Dr. Walsh blasted this type of loose evidence typical of the anti-premature-burial propaganda: "Any man who advanced in proof of a scientific proposition hearsay evidence of what happened three generations ago would at once become a general laughing-stock."

Those heavy blows for ever close
The sun from o'er my head!
The dungeon is my chamber now
My coffin is my bed!
I'm dead unto the living world,
And living with the dead!

NOT THE LEAST important part of the work of the London Society for the Prevention of Premature Burial was public agitation to spread the word about the perils of premature burial. The book *Premature Burial and How It May Be Prevented* was a mainstay in this campaign, and a second edition, edited by Dr. Walter R. Hadwen, was published in 1905. By this time, William Tebb himself was getting old, and most other leading members of the London Society for the Prevention of Premature Burial were timid, fearful of ridicule, and unaccustomed to public speaking. Not so the Swedish author and controversialist Emilie Louise Lind-af-Hageby, who had moved to London in 1905. She came from a well-known noble family, but had grown quite notorious in Sweden as a radical antivivisectionist and suffragette already in her twenties. Young, vigorous, and possessed of independent wealth as well as a burning hatred of the wrongdoings of medical science and the horrors of premature burial, she immediately became one of the society's most active members. When she spoke at its ninth annual meeting, held at the Frascati restaurant in Oxford Street, she was dismayed to find the meeting so scantily attended. She boasted that in Sweden she had held several lectures of her own about the horrors of vivisection and the torments of prematurely buried people. In her sensationalist agitation, she had claimed that tens of thousands of people were buried alive every year, and quoted Franz Hartmann's disgusting case reports at length. After one lecture, at the Agricultural Hall in Stockholm, doctors in the audience had threatened her with a lawsuit and said that she ought to be in prison for daring to introduce such sensational and untrue stories.

Miss Lind-af-Hageby immediately began her one-woman campaign against the perils of premature burial and set up an elaborate lecture tour around the British Isles. This was observed with dismay by several senior clergymen, who wanted the ghost of premature burial to be laid to rest once and for all. She was even barred from some ecclesiastical conferences. On other occasions, the clergymen sent unkind letters to the local newspapers, one of which published a cartoon that depicted her sitting in a coffin, with the caption "The Lady who wants to add to the Terrors of Death!" On other occasions, they arranged for some shrewd clerical debater to attend the meeting to present the establishment's point of view. An experienced controversialist, Miss Lind-af-Hageby was more than a match for these hecklers, however. Once, in Bristol, a clergyman had asked her whether it was really such a good idea to keep corpses above ground for weeks, waiting for signs of putrefaction—surely this would cause far from pleasant smells in the neighborhood. With a discourteous reference to the English people's reluctance to wash, Miss Lind-af-Hageby replied that since she had come to this country, she had learned that some living people smelled far worse than the dead. In Margate, a young man heckled her at length and denied that any case of premature burial had ever taken place.

"Do you have a brain?" asked the irate lady.

"Of course I have" was the puzzled reply.

"In that case, it is as well hidden in your thick skull as many a case of premature burial in the earth!"

When another Margate townsman complained that his children and servants had been scared out of their minds by horrid pamphlets she had put in their mailbox, she replied that they should in fact prefer these pamphlets to the traditional arms of the militant suffragette: two pints of petrol and a lighted match![19] The anti-premature-burial movement was much the poorer when, in 1909, Miss Lind-af-Hageby decided to devote her time wholly to antivivisection propagandism.[20]

In 1910, Miss Ellen Oakes took over Miss Lind-af-Hageby's role of traveling propagandist, but with far less success. After a meeting in Cambridge, during which she was severely ridiculed by some clergymen, she was so mortified that she fell into a decline, just like the heroines of her horrid stories of premature burial, and was not fit for fight until half a year later. She spoke again at the Ecclesiastical Arts Exhibition in Middlesbrough, but although the clergy was not as rude as in Cambridge, she had to endure much ridicule. In 1913, she again collapsed under the strain of arranging, amid the most trying circumstances, an anti-premature-burial stall at an ecclesiastical exhibition in Southampton; later, she decided to withdraw permanently from active duty as a propagandist.[21]

Oh horrible! Oh damnable!
For now with drownèd groans
With frenzy in my stifled soul
That thaws my stiffened bones
I, bursting forth the coffin's sides,
Fall bleeding to the stones.

ANOTHER MAINSTAY IN the activities of the Society for the Prevention of Premature Burial was its journal, the *Burial Reformer*, begun in 1905. At first, this journal, edited by Mr. Arthur Hallam, was quite matter-of-fact: it contained reports of the society's meetings, articles on burial reform by its leading members, and flippant stories of premature burials or lucky escapes ("A corpse asks for beer" or "A corpse jumps out of his coffin") culled from newspapers all over the world. Much was made of "The Accrington Sensation" in 1905. Mrs. Elizabeth Holden had been laid out for dead, and might well have been buried, had not an undertaker noticed the twitching of an eyelid: "pale, wan, extremely weak, she feebly lisped out to a representative of the *Manchester Courier* her recollections of her terrible experience." Dr. Franz Hartmann contributed some gruesome Continental cases: an Italian woman had given birth to a child inside her coffin, and a miner in Budapest had heard noises emanating from the tomb of a newly buried mountain man, who was dug out alive, but "in an undescribable state." The aforementioned James R. Williamson suspected that anesthesia might cause a particular kind of deep trance and that many who died after operations were in fact buried alive.[22]

The society had quite a following among senior military men: General J. M. Earle and Lieutenant General A. Phelps were both active campaigners, and General John P. Hawkins of the U.S. Army was another member. The dean of Llandaff was the society's leading clerical member, seconded by Archdeacon Colley, who claimed that he himself had nearly been buried alive as a child. Dr. Stenson Hooker and Dr. Walter Hadwen were the leading medical members. Hooker had received a considerable number of letters from people saying that after reading Tebb and Vollum's book, they had a great horror of premature burial. At a meeting in 1906, he dryly added that they must have an even greater horror of paying a small sum to avoid this dreadful catastrophe, since he had not been able to coax a single one of them into joining the ranks of the society.[23] Hadwen was a controversial Gloucestershire physician, whose antivaccinationist activities had nearly got him struck off the medical register.[24] The main purpose of the society, to bring about burial reform and the erection of waiting mortuaries, was aided by two parliamentarians in its ranks: Sir Walter Foster and Mr. George Greenwood. The former presented a private bill for measures to safeguard against premature burial in 1908, but without success. He deplored that whereas a bill that aimed to build more battleships had been approved in just three days, his own struggle against the specter of premature burial had gone nowhere in fifteen years: measures to destroy lives were dealt with expediently, but a bill that aimed to save lives was neglected.[25] The *Punch* of February 12, 1908, reported that when a newly elected member of Parliament was asked whether he would support Sir Walter Foster's bill, he guardedly replied that he wanted both sides discussed before he made his mind up: "This is the first we have heard of a Party in favour of Premature Burial."

Active foreign members of the Society for the Prevention of Premature Burial included Dr. Franz Hartmann, of Florence, and our old friend Count Michel de Karnice-Karnicki, who now resided in Viezeggio, Italy. That veteran campaigner Dr. Séverin Icard, of Marseilles, was actually its vice president. Another leading French member was Dr. J. B. Géniesse, who had translated into French a treatise on apparent death by the Spanish Jesuit priest Juan Ferreres, with copious additions and comments of his own. Some of Géniesse's additions were culled from Tebb and Vollum's book; others provide an interesting overview of some scarce French books and pamphlets on the subject.[26] The pope personally congratulated Dr. Géniesse for his philanthropic zeal in the service of the apparently dead. A wealthy Italian nobleman, the marchese Vitelleschi, was so alarmed by the book's horrid stories that he ordered a private waiting mortuary, the Vitelleschi Asylum for Cases of Doubtful Death, to be built in Rome.

After 1910, the Society for the Prevention of Premature Burial fell into terminal decline. The members were few and elderly, the activity low, and the journal, which had been renamed *Perils of Premature Burial* in 1909, became increasingly ludicrous. According to an article from an evening newspaper, a recently deceased workman's wife in Dux, Bohemia, had always had a great fear of premature burial. Her husband was a devout Christian who believed in bodily resurrection, however, and it would have broken his heart if his wife's corpse had been cremated or left to putrefy in a waiting mortuary. As a compromise, she was buried with a crucifix in one hand and a loaded revolver in the other; it is to be hoped that, if her religious husband's notions were correct, she devoutly knelt to Saint Peter behind her crucifix, instead of producing the gun and demanding entry! At Kolpino, near St. Petersburg, a cry was heard when the coffin of a child was put into the earth; inside was a tiny baby calmly sucking a feeding bottle. The sad tale of the young Italian tenor Attioli, who had shot himself for the love of a countess, was also discussed. One of the many thousands of women who attended the funeral, claimed that she had heard the amorous singer sigh in his coffin. An even taller story, from the *Battle Creek Daily Journal*, describes how Mr. Frederick J. Harvey, one of the wealthiest men in Kansas, was buried alive in the family vault from early January until the middle of May. His sweetheart, Miss Lily Godfrey, suspected that he was not really dead, and found him alive in the vault. Since he was in a state of trance, the newspaper learnedly and confidently explains, he needed neither food nor drink; the lengthy incarceration in the coffin also cured his consumption, thus enabling the wealthy American to marry Miss Godfrey and set out on his honeymoon on September 5, 1906. A happy ending indeed! Another ludicrous story tells the sad tale of Mlle Sviscnzadi, of Szilagysombyo (!). She was engaged to marry a young man named Franz Kasinski, but was terrorized by her former lover, who even took a shot at her. She swooned and was declared dead. When he saw his fiancée's lifeless body, Kasinski shot himself dead with his revolver. The woman's "corpse" then sat up with a scream, waved her arms inside the shroud, and tried to get out of the coffin. When told that Kasinski was dead, she collapsed again, and was "now critically ill."[27] A similarly telling contribution, with regard both to the lack of newsworthy material that must have plagued the editor of the *Perils of Premature Burial* and to the intellectual standard of its readership, was submitted by an American lady, Miss B. E. R. Thomson of Brattleboro, Vermont, who was living in a retreat for the mentally disturbed. She claimed that she had been persecuted by a gang of Catholic priests, who had tried to convert her to Catholicism by alternately confusing and bullying her; after she had taken a shot at one of them with a revolver, the cunning papists had used this as an excuse to put her into an asylum. Her interest in preventing premature burials had been made much of when she was declared insane, and had been ridiculed in court.[28]

In 1914, the long-suffering editor Arthur Hallam had to announce that *Perils of Premature Burial* magazine had published its final issue, since the print run had become precariously small. The society struggled on, however, helped by its president during the 1920s, Sir George Greenwood, who devoted much time to the subject of burial reform. The last notice of its activities is from 1936, when the few remaining members merged with another group of enthusiasts, the Council for the Disposition of the Dead.[29]

E'en now contagion and death
Their reeking corpses breathe;
Again I climb their putrid clay
Again it sinks beneath!
And 'midst their musty bones I lie
And gnash my desperate teeth!

THE ANTI-PREMATURE-BURIAL movement of the 1890s should be viewed as one of several internally linked organizations opposed to medical science. It is curious that this movement had its strongholds in Britain and the United States, countries that had been largely spared from the mid-nineteenth-century preoccupation with apparent death and premature burial. A likely explanation is the strong link with organized spiritualism, which flourished in the English-speaking countries at this time. Although some of the members flaunted their political liberalism or radicalism, their aims were basically reactionary, preferring cleanliness and godliness as the key to human health, rather than the newfangled discoveries of bacteria, germs, and vaccines. William Tebb and Alexander Wilder both rejected the bacterial origin of disease. Both also had an interest in occultism and spiritualism, as had Franz Hartmann and Arthur Lovell. Many of the anti-premature-burial agitators, none more than Walter Hadwen and Lizzy Lind-af-Hageby, also opposed vivisection, and some took a stance against medical laboratory research as a whole and against the social and professional power of doctors. There were links also to the proponents of teetotalism and vegetarianism, to the organized spiritualists and quacks, and to the suffragettes and proponents of rational female dress. In London, the *Herald of Health* magazine was conducted by Mrs. C. Leigh Hunt Wallace, author of *Physianthropy; or, The Home Cure and Eradication of Disease*. In 1895, this magazine had articles like "Bad Air as Cause of Disease," "Vaccination, the Delusion of the Century," "Anti-Corset League's Meeting Transactions," and "How the State May Prevent Premature Burial"; the last contained the usual horror stories, and the usual appeal to build deadhouses everywhere.[30]

The criticism of medical authority from these antiscientific reactionaries did have its beneficial points: they rightly pointed out that the unnecessary cruelty practiced by many nineteenth-century vivisectors had no place in modern society, and thus helped raise people's consciousness about the rights of animals. On the other hand, the antivaccination and antibacteriology campaign of Tebb and others was dangerous as well as completely wrongheaded; had they been taken seriously, a full-scale medical disaster would have resulted. The anti-premature-burial propaganda fell in between these two extremes. It must be admitted that the Society for the Prevention of Premature Burial had a point when it attacked the British Burial Act of 1900, which permitted doctors to sign death certificates without having seen or examined the patients. In contrast to those in France and Germany, where a relatively modern system of death certification had been operational for many years, the authorities in both Britain and the United States viewed these certificates from the point of view of demography and population statistics, rather than as a safeguard that the person buried was really dead. On the other hand, the anti-premature-burial propagandists asserted many untruths and recirculated (or in some cases invented) the most loathsome horror stories in their periodicals, catering to the most unreasonable fear of premature burial. In an article blasting the society, a medical journal rightly concluded, "The life of man is already surrounded by a sufficiency of trials and difficulties without adding gratuitous terrors to the list."[31] It is curious that the fear of premature burial, which originated with the works of Bruhier and other eighteenth-century physicians, was to end up as a vehicle for opposition against medical science.

But what happened to the prematurely buried woman whose "drownèd groans" and desperately gnashing teeth have entertained us throughout this chapter? Well, it is the old story again. A gang of grave robbers lowered a rope into the vault where she lay groveling in a heap of putrefying corpses, and with the agility of a circus acrobat she took her chance to escape:

I madly scaled the quivering rope
  And frantic with delight
Rose up from the sepulchre
  A ghastly thing in white
That petrified the robbers' eyes
  With yells of fear and fright.

X

Literary Premature Burials

> How had I fallen into a chasm that closed
> Its dark inevitable arms, and crush't
> Me, bruised and blind! I struck, and struck, and beat
> With bleeding strength, in vain. A hundred hands
> Fought in the gloom with mine as water weak
> At every step there stirred some hissing snake
> I felt as one that's bound, and buried alive
> The black, dank death-mould stampt down overhead
> And cried, and cried, and cried, but no help came.
>
> —*Gerald Massey,* Only a Dream

The earliest literary use of the premature burial theme follows a familiar theme: that of the young lady who is accidentally buried alive but saved by her lover. In his *Decameron,* written in the 1350s, Giovanni Boccaccio describes the attempts of a lecher named Messer Gentil da Carisenda to seduce Catharina, the beautiful wife of a gentleman named Nicoluccio Caccianimico, but he is not given any encouragement by her. She dies some time later, and Gentil decides to visit the tomb "to steal some innocent Kisses"; he reasons that now when she is dead, she can no longer resist his advances. He breaks into the vault and takes liberties with the cold, still corpse. Some days later, he comes back "to touch her Breasts." He is thus employed "for some considerable time," when he suddenly detects that her heart is still beating. Gentil stops his necrophilic activities and rushes her back to his mother's house; the virtuous lady recovers from her death trance and is later reunited with her husband. Another of Boccaccio's more risqué tales also deals with the misadventures of a living corpse. A clever abbot wants to seduce the wife of a stupid, wealthy farmer named Ferondo, but the farmer knows his wife's flighty disposition and guards her jealously. The abbot then deliberately poisons Ferondo with a powder that put him into a deep trance. He is declared dead and buried in a vault, but the abbot clandestinely kidnaps his lifeless body and brings it to a torture chamber, where the luckless farmer is whipped and tortured to make him believe that he is in purgatory for the great sin of being too jealous. The abbot and Ferondo's wife "pass'd the time very agreeably together" until the wife becomes pregnant. The abbot then again poisons Ferondo with his drug, puts him back into the vault, and has him rescued from his premature tomb the next day. At first, everybody thinks that he is a ghost, but Ferondo explains that he was in purgatory and that God spared his life after he promised never to doubt his wife's virtue again. The abbot and Ferondo's wayward wife later "met together as often as they conveniently could, making themselves very merry at the stupidity of the Husband."[1]

The sixteenth-century author Matteo Bandello also used the theme of the lady buried alive, in one of his *Nouvelles*.[2] But while the hero of Boccaccio's tale has been a lecherous fellow, Bandello's two young lovers are pure, innocent teenagers. Gerardo and Hélène are children of two wealthy Venetian burghers, whose palaces face each other across the Grand Canal. After the death of Hélène's mother, she is brought up by a wet nurse, who aids and abets the two young lovers; she even secretly marries them in a mock ceremony, kneeling in front of a statue of the Virgin Mary. But then things go badly wrong for Gerardo and Hélène: he is sent abroad to take care of his family's business, and her father decides that she is to be married to a certain gentleman, without consulting her wishes on the subject. Poor Hélène falls into a decline when she hears this disastrous news. The day before her wedding, the lethargy deepens, and a doctor declares her dead from a "subtle catarrh that has traveled from the head to the heart." She is buried in the family vault. The same evening, the distraught Gerardo goes there to see her one final time. He finds her alive and takes her to his mother's house, where she is nursed back to health. Her presumptive husband challenges Gerardo to a duel when he finds out that his wife is still alive, but the matter is put before the magistrates of Venice. The ubiquitous wet nurse appears, however, to tell them about the secret marriage, and all ends happily. Bandello was probably inspired by an earlier variant of the story, by his fellow Italian Luigi da Porto, but there were several other variants, one of them the sad tale of Romeo and Juliet that was adapted by William Shakespeare.

What is likely to be the earliest premature burial scene in English literature takes place in Thomas Amory's *John Buncle*.[3] In this long and slightly ridiculous novel, the passionate Buncle wanders through the dangerous moors and mountains of northern England, where he encounters various beautiful and talented women, all of whom he marries after successive accidental deaths. The narrative is generously interspersed with eloquent digressions on religious and literary subjects, and Rabelaisian descriptions of excessive eating and drinking. After the death of Miss Spence, who is felicitously described as having "the head of Aristotle, the heart of a primitive Christian, and the form of Venus de Medici," with whom Buncle had discussed calculus after dinner, he looks for yet another wife. He abducts Miss Agnes Dunk, the daughter of a wealthy miser, and she agrees to marry him. Just a few days after the wedding, she dies of a fever, and the grieving Buncle buries her after keeping the body seven days above ground. A certain Dr. Stanvil rifles the tomb and steals the body, which he wants to use for anatomical research. But when the anatomist's knife scratches her at the beginning of the *linea alba*, as Buncle tactfully expresses it, Miss Agnes awakes. She is nursed back to health by the amorous doctor, who later marries her. *John Buncle* is not a little surprised when he meets his "dead" wife in the guise of Mrs. Stanvil; it is characteristic of his jolly, carefree attitude that he has no qualms of conscience for having buried her alive. Instead, he edifies the reader with a short discourse about the risks of premature burial. *John Buncle* was published in 1756, a few years after Bruhier's book on the uncertainty of the signs of death had appeared in English translation, and some of the cases given by Amory are clearly repeated from this popular and widely read book.

In the late eighteenth-century gothic novel, premature burial was not infrequently used as a literary motif.[4] In Mrs. Showes's *The Restless Matron*, a lady is deliberately buried alive by her wicked husband.[5] She thus recounted her plight to the novel's heroine:

> "It was midnight; I heard the clock strike twelve. I opened my eyes, and raised my hand: the impenetrable darkness I was in, terrified me . . . I raised myself up, and looked around me: I was in a shroud; a coffin, that was placed in the midst of innumerable others, was my bed; a silver lamp, that afforded a faint glimmering of light, was suspended by a chain in the midst of the vault—I was buried alive!"
>
> This account was too dreadful for Agnes's weak spirits to bear; she gave a faint scream, and sinking back upon the couch, fainted.

As could be expected, the premature burial theme also shows up in German eighteenth-century literature. The sentimental novelist Jean Paul Richter, who had quite a large following during this time, was one of the first to let a character fall victim to the uncertainty of the signs of death. In his novel *Siebenkäs*, published in 1796, the main character, a young lawyer, plays dead to get away from his wife. The doctor is completely fooled by this simple trick, and Herr Siebenkäs is declared dead, put in a coffin, and buried, to be dug up and rescued by a friend in the next chapter.[6] There were quite a few other sentimental German stories about apparently dead people being rescued by philanthropists, many of them published in the 1790s or early 1800s. Poetry treated this subject as well. The popular collection *Des Knaben Wunderhorn* contains a very sentimental poem entitled *"Der Scheintod,"* about some children walking to the churchyard to put some flowers on the grave of their mother, who had recently died during pregnancy. They are mystified to hear noises emanating from her nine-day-old grave—not screams of horror or the scratching of broken fingernails, but a clear voice singing a lullaby. The children run home to tell their father, and he comes with them to the churchyard and also hears the voice. In a frenzy, he drags the gravestone away and breaks the coffin open; inside is his wife, alive and well, holding a smiling infant she had given birth to underground.[7] The buried-alive theme also appears in the Scandinavian literature. Probably the first to use it was the Danish poet Adam Oehlenschläger, who wrote a short story entitled "Reichmuth von Aducht" after visiting Cologne in 1813.[8]

THE WRITER WITH the most premature burials per page must be Edgar Allan Poe, whose unwholesome fascination with this subject is apparent to every devotee of his horror stories.[9] It has been questioned whether he himself lived in fear of being buried alive or merely wanted to use a well-known and horrific subject to curdle the blood of his readers. In the 1830s and 1840s, it was neither uncommon nor abnormal to be concerned about the risk of being buried alive; indeed, some of the leading European medical authorities on the subject were of the opinion that live burials were common. The popular newspapers and magazines, like *Blackwood's Edinburgh Magazine*, the *Casket*, the *Southern Literary Messenger*, and the *New-York Mirror*, frequently published stories of premature interment—some of them plainly fictional, others claiming a factual origin. It is likely that these tales provided Poe with his background reading on the subject, particularly as his various editorial positions required him to keep abreast of periodical writing.[10]

In Poe's story "The Premature Burial," the narrator lists some instances of premature interments to persuade the reader that this phenomenon is a dreadful reality. One of them is a garbled version of the Two Young Lovers legend, telling how Mademoiselle Victorine Lafourcade is reunited with her lover, the poor litterateur Julien Bossuet, after he has opened her tomb to cut off some curls of her hair. The story is said to have occurred in Paris in 1810, but the ending is the traditional one: her wicked banker husband recognizes her, and the case is brought before a "judicial tribunal." It has been shown that Poe must have read the tale of Mademoiselle Lafourcade in a Philadelphia periodical.[11] The second story is about the Englishman Edward Stapleton, who is buried alive while in a state of trance, but rescued by body snatchers who bring his body to a private laboratory. Before dissection, they decide to stimulate the corpse with a galvanic battery, and Stapleton awakens from his trance. This story is derived from an article in *Blackwood's Magazine,* and the name Edward Stapleton is probably Poe's own invention.[12] Poe's third "factual" instance is a horrific one, and brilliantly described. The wife of an eminent citizen in Baltimore was buried in the family vault. Three years later, the vault is reopened, but when the husband swings the door open, the skeleton of his wife, dressed in her shroud, falls rattling into his arms. She had broken out of her coffin, but been unable to break through the vault door; she died a second death leaning toward it, and after her shroud became entangled in some ironwork, rotted in an erect position. This tale can also be traced to a fictional story, published in the *Southern Literary Messenger* of 1837, six months after Poe left an editorial position with this magazine.[13] It is instructive that while the lesser writer, a certain James F. Otis, describes, in lurid detail, how the wretched woman tears off her shroud and screams in agony during the desperate struggle, Poe leaves it to the reader's imagination to supply the minutiae of what occurred in the vault.

The protagonist of Poe's "The Premature Burial" is a young man who suffers from a strange neurological disease, which makes him

The famous illustration for Poe's "The Premature Burial" by Harry Clarke. *Reproduced by permission of Jennie Gray, The Gargoyle Press (Chislehurst, UK).*

prone to epileptic seizures, after which he often lapses into a state of deep lethargy. He is an avid reader of books about apparently dead people awakening in the tomb and suffering the most dreadful tortures. As a precaution, he buys a security coffin, and he always carries with him written instructions that he is under no circumstances to be hastily buried, should he suddenly collapse in the street. Poe may have read an article in the *Columbian Lady's and Gentleman's Magazine* about a luxurious "Life-preserving Coffin" exhibited at the American Institute in New York. Intended for vault burials, it was equipped with springs and levers on the inside, which would make the lid fly open at the slightest movement of its occupant.[14] The conclusion of the story is that he awakes in absolute darkness, lying in a musty-smelling, narrow area, surrounded with wooden walls. His jaws are bound up like those of a corpse; with horror, he understands that he has been buried alive. In vain, he searches for his safety equipment inside the coffin, but he finds nothing except the unyielding walls of his narrow house. He realizes that he must have been buried by strangers and "nailed up in some common coffin—and thrust, deep, deep, and for ever, into some ordinary and nameless *grave.*" His wild cry of terror and anguish is heard by some sailors, however, and it turns out that he has really been lying in a narrow bunk on board a trawler; the musty smell came from the ship's cargo, and the cloth around his jaws was his own necktie, which he had been using as a nightcap. This nightmare experience made him forget his fears of being prematurely buried, and transformed him into a saner and healthier man: he burned his medical books on the uncertainty of the signs of death and read "no *Night Thoughts*—no fustian about churchyards—no bugaboo tales—*such as this.*"

"The Premature Burial" is not one of Poe's better stories; it is marred by the contrast between the rising tension of the story and the tongue-in-cheek quality of the ludicrous ending. This did not prevent it from being taken seriously by at least one credulous American magazine writer on premature burial, who quoted its case reports with approval. "The Premature Burial" is by no means the only one of Poe's tales that use the theme of a premature burial. In "The Cask of Amontillado," the wretched Fortunato is deliberately walled in by his enemy Montresor, and the "Black Cat" accidentally suffers the same fate, only to make its presence in the premature tomb known by a ghostly caterwauling. In several of his stories, like "Morella" and "Ligeia," the female protagonists supposedly die from diffuse nervous afflictions, but return from the tomb. In "The Fall of the House of Usher," Lady Madeline Usher dies from an obscure cataleptic disorder. Her brother Roderick does not want her to be hastily buried in a remote churchyard, but orders to have her coffin entombed in a deep underground vault. She revives from her trance, however, breaks out of the vault, and reappears, like the Angel of Death, to reclaim her wretched brother. It is left undecided whether Roderick knew all the time that his sister was buried alive, and exactly what the relation had been between these two siblings living alone in their rambling castle. Lady Madeline's return from the tomb is an event quite different from the rescue of the passive sleeping beauty in the Lady with the Ring or Two Young Lovers legends; she breaks out of the vault by using brute force and returns to haunt the man who consigned her to the grave. This is reminiscent of another popular tale, which can be found in more than one nineteenth-century French book on premature burial: that of the termagant wife escaping from the tomb.[15] The original story is about a meek French clerk bullied by his strong, forceful wife. One day, the wife dies, so it is believed, and the pallbearers carry her coffin away to the funeral. But they drop the coffin while negotiating a narrow street corner, and the wife revives. Furious, she walks back home in her shroud and scolds the wretched husband for consigning her to a premature tomb. When the wife really dies, some years later, the husband tells the pallbearers, "I implore you, messieurs, be very careful when turning corners!"

It is also possible to reinterpret another of Poe's more powerful stories, "Berenice," by means of knowledge about the old European legends on premature burial, which had been assimilated into nineteenth-century magazine and newspaper fiction. Egaeus, the narrator, is a sad, bookish loner, who lives in his own fantasy world. He suffers from an obscure species of monomania, which makes him intensely interested in observing and contemplating the most ordinary objects. For example, he might spend an inordinate time examining some detail in the typography of a book. The only other living member of Egaeus's family is his cousin Berenice. Once she was a gay, vivacious girl, but a variety of severe maladies, among them a kind of epilepsy that might terminate in a deathlike trance, have left her a frail invalid. One day, the gloomy Egaeus contemplates Berenice's emaciated frame and wasted beauty. He becomes morbidly fascinated with the spectacle of her teeth and thinks of them night and day. The symbolism of this is unclear, but it has been speculated that the teeth represent the incorruptible part of Berenice's face, which will remain unchanged even long after death.[16] Two days later, Egaeus's servants tell him that Berenice has died, and she is buried the same night. In the final scene, Egaeus sits reading in his library, when a servant comes in to tell him "of a wild cry disturbing the silence of the night—of the gathering together of the household—of a search in the direction of the sound;—and then his tones grew thrillingly distinct as he whispered me of a violated grave—of a disfigured body enshrouded, yet still breathing, still palpitating, *still alive*!" The obtuse Egaeus does not immediately react to this compelling announcement, but calmly remains seated in his reading chair. The horrified servant then points to Egaeus's muddy clothes, stained with gore, his hand "indented with the impress of human nails," and a spade that has been left standing against the wall. Egaeus gives a shriek and grabs a little box on the table beside him. It falls and breaks into pieces, and he sees "some instruments of dental surgery, intermingled with thirty-two small, white and ivory-looking substances that were scattered to and fro about the floor." The demented Egaeus had gone to Berenice's grave, dug up the coffin, and forcibly extracted all her teeth; it had not deterred him that his cousin had awakened from her trance during this diabolical dentistry. If Berenice is the Lady with the Ring, Egaeus is the grave robber who mutilates her, or rather, from the motives that drive him to his perverted actions, the Lecherous Monk. Several versions of the former story were current in the contemporary magazine fiction, but their happy endings, with the Lady safely returned to her doting husband, are in sharp contrast to Poe's morbid, blood-spattered fantasy.

An illustration for Poe's "Berenice" by the German artist Alfred Kabin. *From the author's collection.*

EDGAR ALLAN POE'S obsession with premature burial was mirrored, although in a distinctly strange manner, in the eccentric literary productions of the learned German lady Friederike Kempner. The daughter of a wealthy Jewish landowner, she enjoyed a privileged upbringing and could study the German classics without any thought of one day having to earn her living. At an early age, she had chanced to read a horrid pamphlet about premature burial, and this experience marked her for life. In 1853, when she was just seventeen years old, she wrote a pamphlet (or *Denkschrift*) of her own, containing a passionate appeal to the German states to build waiting mortuaries to save the apparently dead citizens from the perils of a premature burial.[17] She boldly claimed that a person could be in a state of trance for twenty days without any signs of putrefaction, and that listening to the heart with a stethoscope was completely useless, since an individual in a death trance could lack an audible heartbeat. Nor had she any scruples about giving the most lurid case reports, which were mainly unreferenced and unverified, in the true tradition of the anti-premature-burial propagandist. Amazingly enough, she even abstracted the examples from Poe's "The Premature Burial" and presented them as true instances of premature interment. Another example came from closer to home. When a grave was dug in the cemetery of Prosenitz in Mähren, a coffin was found to contain a still not decomposed corpse in a horrible, unnatural position, with the teeth sunk into the upper arm. It was the body of a postmistress who had been hastily buried a few years earlier. The prematurely buried woman's doctor blamed himself so much for this horrid disaster that the same night, he himself died of the cholera. Another of Friederike's most pleasant stories concerns a noble countess, who was buried in a vault situated directly under a church. After the funeral, a knocking sound could be clearly heard, coming from inside the vault. The verger thought it was a ghost, and the parson shut the church doors and forbade him to tell anyone, since the church might receive complaints if "mistakes" like this one were not left underground. The horrid sound continued for seven days, but then ceased, to the relief of the inhuman parson. When the vault was opened many years later, a skeleton was found on its stairs, and the countess's coffin was empty. Here is a sample of Friederike's overblown tirades:

> Is there a more horrific, more hopeless fate than that of being *buried alive*? . . . the person who awakes in the coffin has been repudiated underground by his own fellow beings. . . . This immeasurable, terrifying scourge of humanity must finally be conquered, and no more of our people fall victim to the most hideous, desperate fate, the *hor-*

> *rid struggle* in the *coffin*; this is a *misery* that lacks a *name*, since it is mute: the *groans* of its *hapless victims* underground can *neither be heard nor seen*!

In spite of its gruesome subject and linguistic oddities, Friederike Kempner's *Denkschrift* became quite a best-seller. In 1857, its proud author could announce the fifth, revised edition, which was translated into Swedish the same year. By this time, Friederike had followed Hufeland's noble example: she had personally supervised the erection of a *Leichenhaus* on the family estate. She sent free copies of her pamphlet to many princes and magnates all over Europe, and some of them, like Duke Eugene of Württemberg, Prince Friedrich Wilhelm of Prussia, and Emperor Napoleon III of France, sent her letters of approval for her humanitarian zeal. Countess Humiecka built a waiting mortuary on her estate in Posen after reading the *Denkschrift*. Prince Anatol Demidoff had the *Denkschrift* translated into French, and also built a waiting mortuary on his estates in Geneva. Friederike Kempner directed a steady bombardment of letters, poems, and gruesome case reports toward King Wilhelm of Prussia, who was kind enough sometimes to reply to her missives. At the onset of the Franco-Prussian War in 1870, she requested permission from her royal correspondent to be allowed to come to the battlefront in order to examine the corpses of the brave German soldiers, to make sure they were really dead before they were buried. This humanitarian mission was not allowed, however, although the patriotic Friederike stated that she had no similar qualms about prematurely buried French soldiers. She had to be content with erecting a military hospital on her estate, in which she personally cared for the wounded. In between her duties as a nurse, she still found time to warn King Wilhelm about the risk of his brave warriors being buried alive; among her missives was a ludicrous poem:

> *In the German race's*
> *Fight for land to live in*

These pale, heroic faces
Of the dead—may still be living!

Great Emperor, with France enslaved
Remember that it's surely
A dismal end to soldiers brave
To be buried prematurely!

In March 1871, maybe as a result of Friederike Kempner's agitation, the newly crowned Emperor Wilhelm I mandated an obligatory five-day waiting period between the time of death and burial in the entire German Empire. Although her interest in premature burial and *Leichenhäuser* was a lifelong one, the learned lady now aimed some of her philanthropic activities at other injustices: she campaigned against solitary confinement of prisoners, antisemitism, and vivisection. The success of the *Denkschrift* went to her head, and Friederike Kempner began to fancy herself a celebrated author and poetess. A stockily built, prim spinster who bore some resemblance to an elderly Queen Victoria already in her thirties, she was a prolific writer of pathetic, overblown poems, which she herself considered equal to the works of Schiller and Goethe. She had little sense of either grammar or meter, and an almost uncanny ability to just miss the mark in a poem, or to break the tone at the last moment. She published her poems privately, but solicited reviews with characteristic determination, sending books to crowned heads all over Europe, as well as to influential newspapers and literary critics. One of the latter, the editor Paul Lindau, was amused by Friederike's linguistic equilibristics and recommended her poems highly; although other critics complained that she brutalized their noble Teutonic mother tongue, the reading public found her poems exquisitely funny. She often returned to her own pet subject of premature burial, as in "Der Scheintodte":

And he slept and slept so long
That nothing could make him come round;

Friederike Kempner. *From a photograph in the author's collection.*

But when buried with flowers and song,
The "corpse" stretched himself underground.

Another favorite was "The Prematurely Buried Child," in which the stirring outcry of the agitated soldier, which so efficiently dispels the graveyard gloom induced by the earlier stanza, echoes Friederike's own interest in burial reform:

A black and cloudy night,
The graveyard's dark and still;
Under earth and out of sight,
A coffin'd child awakes
And gathers his weak will.

He tears the shroud and hood
With fingers black and blue

He fumbles—but just wood
For every move he makes
"Mummy, where are you!?"

His bloody hands they knock
Unyielding coffin walls
Half dead with fright and shock
"Hear, I am not dead!"
But no one heeds his call.

"Will I live forever in here?
Ach God! Just bring a light!
Farewell now, mother dear!"
With a final shudder of dread,
He faints from pain and fright.

In his coffin, all alone
At length the poor child dies
And gives his final groan.
Above the cross of stone
A weeping angel flies.

. . . .

A German soldier bold and brave
Hears how this boy died in his grave.
He could look danger in the eyes
But now he trembles, moans and cries.

"A house for the dead! A house for the dead!"
He shrieks in a voice full of horror and dread,
"We do not want, from fickle chance
To be buried in a state of trance!"

"Yes! 'Tis a house for the dead that we crave
to save us for dying within *the grave*
And if you are not blind, or quite off your head,
'Tis silly *to bury us before we are dead!"*

Friederike Kempner's parents were deeply ashamed of their daughter's dubious literary fame. Fearful that not only the Kempner family but the entire Jewish community in Germany might become a laughing-stock after these ludicrous poems had become popular, they secretly purchased what remained of the first edition and destroyed it. The publishers issued new editions, however, and Friederike's fame was such that the family could not keep up with the demand from the readers hungry for amusement.[18] In 1903, the eighth edition was published, but by that time even the naïve Friederike had come to know the true reasons for her fame. In a furious preface, she swore that the day would come when her name would be etched in bronze and the true literary value of her poems fully appreciated. The following year, she died of a stroke. In her will, she requested that an electrical apparatus be installed in her coffin, so that she could ring for help if she woke up, but this device proved superfluous, since she was cremated.

One of Friederike Kempner's contemporaries, the famous Swiss poet Gottfried Keller, completely outdid even her. He devoted an entire cycle of poems to describing the thoughts of a man buried alive in a coffin under six feet of earth.[19] It is a beautiful, although macabre poem and contains some haunting images: the prematurely buried man feels the pine boards of his coffin, and they remind him first of a ship's mast and then of the first pine tree he ever saw, a Christmas tree ablaze with candles. He discovers that he has eaten a rose that had been put between his hands before the coffin lid was screwed shut, and is sad that he will never know whether it was a white or a red rose. The final image is of a surging sea of life, a vast landscape with no horizon; for the prematurely buried man, whose time is rapidly running out, eternity is captured in a single intake of breath. The Swiss composer Othmar Schoeck made Keller's "Lebendig Begraben" into a forty-five-minute long orchestral song cycle, which begins with a realistic impression of earth falling down onto a coffin lid.

There are quite a few examples of well-known literary men who feared a premature burial. The famous German philosopher Arthur Schopenhauer freely admitted to a fear of premature interment and stipulated that his corpse rest above ground five full days before burial. The Austrian writer Johann Nepomuk Nestroy was much more elaborate in his precautions. In his will, he declared that the risk of premature burial was the only thing he feared in his present situation and that his studies of the literature on this subject had taught him that the doctors could not be relied on to distinguish dead people from living ones. His body was to be kept in an open coffin for two days, in a waiting mortuary with a signaling apparatus that would herald any signs of returning life. Even after burial, the coffin lid was not to be nailed shut.[20] Russia's most famous example of the nineteenth-century fear of premature burial is the author Nikolai Gogol. In a letter to a friend, he wrote that it was amazing that, as he had read in a book, the human organism could exist in a state of trance for long periods of time and that the individual could see, hear, and feel, but not move, speak, or do anything to prevent being prematurely buried. In his will, Gogol specified that he not be buried until there were clear signs of putrefaction and until it was ascertained that there was no heartbeat and no peripheral pulsations. It has been speculated that, in spite of his precautions, Gogol was really buried alive. When his coffin was exhumed many years after his death in 1852, it was seen that the body was lying on its side; according to a German author, there was only one explanation for this: Gogol had awakened in the coffin, and a terrible struggle had occurred.[21]

AN IRISH MASTER of the horror story, Joseph Sheridan Le Fanu, a contemporary of Poe's, used the premature-burial theme in his short story "The Room in the Dragon Volant."[22] A wealthy young Englishman falls afoul of a band of French impostors. They plan to steal his money and dupe him to drink poison. He falls into a deep trance, is captured, and put into a coffin. Although paralyzed, he remains conscious and can hear how the rogues plan to have him buried alive. In the nick of time, they are stopped by the French police.

Marie Corelli was one of the most celebrated popular novelists of the late nineteenth century; her books were massive best-sellers, and even Queen Victoria admitted to being an avid reader of them. Corelli's first book, *A Romance of Two Worlds*, was an immediate success; to follow it up, she quickly wrote another high-strung melodrama, tentatively titled *Buried Alive*. It was allegedly based on true events during the plague in Naples of 1884. The book's hero is Count Fabio Romani, a wealthy young Neapolitan nobleman. When he tries to help the poor during a cholera epidemic, he himself catches the disease and is declared dead. The reader of this book will not be surprised at what happens next. Romani awakes in his coffin, which has been put in the family vault: "But what was this that hindered my breathing? Air-air! I must have air! I put up my hands—horror! They struck against a hard opposing substance close above me. Quick as lightning then the truth flashed through my mind! I had been buried—buried alive; this wooden prison that inclosed me was a coffin!"

With remarkable resilience for a cholera victim who has withstood several days of premature burial, Romani breaks out of his coffin "with the strength and frenzy of an infuriated tiger." After tearing a vampire bat from his neck, he wades through the piles of putrefying corpses in search of a way out. The Romani family vault has a rich and varied fauna, and the wretched count has to watch out for its many venomous snakes and reptiles. Suddenly a huge owl swoops down at him, but the sturdy nobleman vanquishes it after a desperate fight. By a coincidence that stretches the imagination even of the reader of a cheap Victorian sensation novel, he then finds not only a convenient exit from the tomb but also a huge treasure deposited there by a gang of robbers. When the resuscitated count returns home, he comes upon his faithless wife in the arms of an old friend of his, Signor Guido Ferrari; it is apparent from their conversation that she never loved Count Fabio and deceived him with Guido even before his supposed death.

Count Fabio Romani vows revenge. His appearance is totally changed by his ordeal in the tomb: his face is pale and haggard, and his hair became white overnight. The wicked Guido and the faithless wife do not recognize the resurrected count when he is reintroduced to them under an assumed name. Count Fabio then seduces his own wife and gloatingly tells Guido what has happened; in the resulting duel, he shoots the rival dead. In the highly charged final scene, Count Fabio takes his wife into the family mausoleum, and tells her his real identity, and that he was buried alive. She tries to stab him, but is crushed to death under a huge boulder that conveniently falls from the roof of the vault. Corelli's publisher, George Bentley, was quite impressed with this overblown melodrama, although he insisted on changing its title to *Vendetta*.[23] Published in 1886, it caused quite a stir: the Prince of Wales requested a copy. It remains one of Corelli's better-known novels.

THE SENSATION NOVELIST Wilkie Collins's work has withstood the test of time much better than Marie Corelli's; he was a writer of considerable talent and imagination, and his books are still widely read. Collins was a nervous, high-strung man and addicted to opium; it is not surprising that he was also yet another victim of the premature burial scare. He particularly distrusted Continental doctors, as did not a few of his xenophobic countrymen. When staying at a hotel abroad, he always put a note on the bedroom mirror, saying that in the case of his (presumed) death, he should not be buried until a competent English doctor had been consulted. Collins very likely visited the waiting mortuary in Frankfurt, which he used as the setting of his 1880 novel *Jezebel's Daughter*.[24] This novel is one of his many potboilers, far inferior to his tensely plotted masterpiece *The Woman in White;* the use of the German *Leichenhaus* is supposed to add a frisson of horror to the narrative, but instead has painful hints of the ludicrous.

The plot of *Jezebel's Daughter* is that a certain Mrs. Wagner has died; her corpse is taken to the deadhouse for three days of supervision before the doctor can sign a certificate that putrefaction has occurred. A foundling boy named Jack, to whom the dead lady had shown much kindness, mourns her bitterly and refuses to leave the body, although the doctor shows him the brass thimbles fitted to the corpse's fingers, the strings, and the rest of the machinery. Together with the old night watchman of the deadhouse, Jack later drinks prodigious amounts of brandy, and they sing a song about a former watchman of the *Leichenhaus* who went mad as a result of his ghoulish occupation:

The moon was shining, cold and bright,
In the Frankfort deadhouse, on New Year's night
And I was the watchman, left alone
While the rest to feast and dance were gone;
I envied them their lot, and cursed my own—
Poor me!

Unbeknownst to either of these sottish, obstreperous fellows, a certain Mme Fontaine had sneaked into the deadhouse; she had in fact poisoned Mrs. Wagner, and wants to gloat at the sight of her remains. But not even the steely nerves of this evil hag are able to withstand the sight of the corpse: "There, grand and still, lay her murderous work. There, ghostly white on the ground of the black robe, were the rigid hands, topped by the hideous machinery which was to betray them, if they trembled under the mysterious return of life." Mme Fontaine seeks comfort with Jack and his inebriated companion, but they jeer and deride her. They sing,

Any company's better than none, I said;
If I can't have the living, I'll have the dead!
In one terrific moment more,
The corpse-bell rang at each cell door!

While they are singing and reveling, the bell in Mrs. Wagner's cell rings time after time, but only Mme Fontaine hears it, since the other two are too drunk. At long last, the apparently dead lady manages to make her resurrection known, and Jack is reunited with his mistress. Mme Fontaine, who inadvertently drinks a glass of her own poison, dies in the agonies considered suitable for a Victorian villainess.

The *Leichenhaus* in Munich is the scene for "A Dying Man's Confession," one of Mark Twain's short stories.[25] The family of a German immigrant in the United States is attacked by two robbers, who kill his wife and child. One of the robbers has no right thumb. The immigrant swears vengeance on them and manages to track one of them down and kill him. He then returns to his own country and becomes night watchman at the Munich waiting mortuary. One evening, he is horrified when the great corpse bell rings; down the outside row of shrouded figures, one is sitting upright and wagging his head from side to side. The German recognizes him as the surviving robber, the one who actually murdered his wife and child. He deliberately withholds the cordials in the *Leichenhaus* pharmacy and himself drinks the brandy intended for the living corpse while taunting his helpless, shrouded enemy and watching him die. The grim conclusion: "It is believed that in all these eighteen years that have elapsed since the institution of the corpse-watch, no shrouded occupant of the Bavarian deadhouses has ever rung its bell. Well, it's a harmless belief. Let it stand at that."

A TIME-HONORED literary use of the apparent-death theme is that some sorcerer or skillful physician possesses a drug capable of inducing a cataleptic trance, indistinguishable from death itself. After being declared dead, the individual who has taken this drug can return from the tomb in a later chapter, to create a dramatic effect. As we saw, this theme was used by Boccaccio; it also occurs in Alexandre Dumas's *The Count of Monte Cristo*. Here the young Valentine de Villefort, the daughter of the wicked procurator who had deceived Edmond Dantès, is gradually poisoned by her murderous stepmother, who had already killed off a considerable part of the family to secure her own son's inheritance. The omnipotent Monte Cristo, who has seen through her nefarious plot, comes up with a clever, if rather far-fetched, plan of his own. With a powerful antidote, he neutralizes the poison every time it is administered, and finally he visits the girl at night to inform her of the impending danger. This time, he replaces the poison with a drug of his own, which is capable of inducing a state of deep trance, with these words: "Whatever may happen, Valentine, do not be alarmed; though you may suffer; though you may lose sight, hearing, consciousness, fear nothing; even if you should awake and be ignorant where you are, still do not fear; even if you should find yourself in a sepulchral vault or a coffin." This overblown tirade is, little wonder, greeted with the cry "Alas! Alas! What a fearful extremity!" from the terrified girl. Everything moves smoothly according to Monte Cristo's plans, however: declared dead by both doctor and coroner, Mlle de Villefort is interred in the family vault. After the Villefort family has been rightfully punished, the distraught count does not forget to have Valentine rescued from the vault and reunited with her fiancé, after she had been buried alive for almost a hundred pages.[26]

The French author Guy de Maupassant admired Edgar Allan Poe, and some of his masterly short stories clearly show Poe's influence. The uncertainty of the signs of death and the horror of a reviving corpse is a recurring theme in some of his more macabre productions. In the short story "Le souris de Schopenhauer," the narrator meets a consumptive German who was once a student of the great German philosopher Arthur Schopenhauer. Although the Frenchman detests Schopenhauer's skeptical and sarcastic ideas, he listens attentively when the German speaks about how his master used to sit at the beer cellar discussing philosophy, smiling his characteristic, horrible smile. After Schopenhauer's death, his devoted students took turns to wake over the body. The corpse appeared remarkably lifelike, the face had its unforgettable smile, and the two students would have expected the great man to stand up any moment, had it not been for the pungent fumes of putrefaction. Suddenly, one of them heard a noise from the room with the corpse, and they decided to investigate. The German student was horrified to see that Schopenhauer was no longer smiling;

instead, the face was contorted into a horrible grimace. He called out, "He is not dead!"—but there was no denying the appalling stench. The ludicrous dénouement is that the other student suddenly pointed to the floor: there lay Schopenhauer's dentures, complete with their springs, which had been catapulted out of the mouth when decomposition had loosened the rigidity of the jaw muscles. It is unknown whether this powerful, macabre story was inspired by some similar mishap during a death vigil in Paris, or whether Maupassant had read some newspaper report of Schopenhauer's last days and his precautions against premature burial.

Another of Maupassant's well-known short stories, "Le tic," shows the influence of Poe even more clearly. At a spa, the narrator meets a Frenchman with prematurely white hair, who has come there with his invalid daughter, Juliette. The Frenchman has a spastic mannerism: every time he wants to reach something, his hand makes a rapid swerving movement before he can take what he wants. The daughter always wears a glove on one hand. Walking with the narrator in the park of the spa, the white-haired Frenchman tells him their strange story. Poor Juliette was suffering from what was thought to be heart attacks, and one day she collapsed and could not be revived. A doctor declared her dead, and the father had her buried in the family vault after a death vigil of two days. She was buried with all the jewelry he had given her, and dressed in her party frock. When returning home after the funeral, the distraught father, half mad with grief, dismissed his old valet Prosper and sat all night in front of the cold empty fireplace, mourning his daughter. Suddenly, there was a loud peal of the doorbell, which echoed through the empty house as if it were a vault. Surprised and fearful, the old man went to open the door. A mysterious white figure stood outside. When the old man asked "Who are you?," it replied, "Father—it's me," and advanced toward him. Convinced that he was seeing a ghost, the terrified Frenchman retreated, trying to ward off the apparition by making a gesture with his hand, which stayed with him as a spastic mannerism ever since. Juliette finally managed to explain that she had been buried alive and that someone had dug up her coffin and then cut off one of her fingers to steal a valuable ring; the bloodflow from the amputation had restored her to life. The father helped her inside and rang the bell for his faithful old valet to light the fires. But when the valet entered the room and saw the girl, he opened his mouth in a gasp of horror and fell backward, stone dead. It was of course he who had robbed and mutilated the corpse, and just as in Maupassant's source, the legend of the Lady with the Ring, the thieving menial met with a both prompt and suitable punishment.[27]

In 1884, at the same time that Maupassant published his short stories, Emile Zola's "La mort d'Olivier Bécaille" appeared.[28] The protagonist of this story is a frail invalid Frenchman, the victim of repeated "nervous attacks." He marries a much younger wife and moves with her to Paris. Not long afterward, he again falls ill, and his wife and landlady believe that he has in fact expired. But just like the traditional victim of the death trance, Olivier is still conscious and can hear and feel everything, although unable to move or speak. His wife and landlady keep vigil over the "corpse," and a doctor is called to write out the death certificate. Olivier awaits the arrival of the medical practitioner with impatience: surely, this well-trained, competent professional must recognize that he is not dead, just a lifeless victim of epilepsy. But poor Olivier's trust in medical science is sadly misplaced: a tired, careless doctor appears and does not even listen for the heartbeat or feel the pulse; he just writes out the death certificate and leaves. Again like the traditional victim of the *Scheintod*, whose torments were described in so many scare-mongering pamphlets, Olivier is unable to make it known that he is still alive, although the preparations for his burial are going on all around him. He is put in a coffin, the lid is nailed down, and he hears the voice of the priest during the funeral, before fainting when the coffin is rudely dumped down into the grave.

When Olivier wakes up, his cataleptic fit is over, and he can move again. But as his hands pass along the wooden boards of the coffin, he remembers his desperate situation: "I screamed, and my voice, reverberating in the pine box, sounded so awful that it terrified me. My God! Was it true, then? I could talk, cry out that I lived, and my voice could not be heard! I was a helpless prisoner under the earth!" He screams, scratches the coffin lid with his nails, and bites his arms in agony. With an almost superhuman effort, the Frenchman then manages to regain his sangfroid. Using a nail extricated from the coffin, he tries to make a deep gash in the coffin lid. Turning over on his stomach and rising to his knees, he is able to split the lid from one end to the other. But his plan to use the lid as some kind of shield while tunneling upward from the grave comes to nothing, since great clods of earth fall down on him, and the board is almost impossible to move. Instead, he tries kicking at the end of the coffin with his heels, and it gives way surprisingly easily. It turns out that another grave has been dug next to his, and Olivier pushes through a thin wall of earth and crawls into it. He is saved! With surprising vigor, considering the ordeal he has just been through, he climbs up from the grave and begins to walk home. He collapses in the street, however, and spends six weeks unconscious at a private hospital.

After being permitted to take a short walk by his doctor, Olivier wants to see what has happened to his young wife. Afraid that she would be killed by the shock of seeing him again, he decides to first seek out his landlady. He overhears the landlady gossiping at the local café, saying that Olivier's death was really all for the best: his poor wife was better off being rid of this dissatisfied, grumbling invalid, and she should definitely marry her new admirer, a younger and more attractive lodger. The philosophical Olivier agrees that he is better off dead, and sets out to travel the world; death no longer terrifies him, and, like the protagonist of Poe's "The Premature Burial," he becomes a saner and healthier man. The two stories also share the first-person narrative and the mysterious cataleptic condition of the victim. But here the similarities end. While Poe's protagonist is basically a figure of fun, whose exaggerated fear of premature burial is ridiculed, there is no hint in Zola's story that Olivier Bécaille has any similar fears, in spite of his alarming medical condition. Apart from Poe, Zola was probably also inspired by one of the many anti-premature-burial pamphlets of the time. The description of Olivier's cataleptic death trance resembles those touted in the pamphlets, and Zola's argument that the wretched man did not suffocate in the coffin—that an individual used almost no oxygen while in a death trance—could almost have been directly quoted from one of these publications, as could the mechanisms given for his awakening in the warm, fertilizing earth.

The Swedish author Selma Lagerlöf was awarded the Nobel Prize in Literature in 1909 for her powerful novels *The Story of Gosta Berling* and *Ring of the Lowenskolds* and for her classic children's book *The Wonderful Adventures of Nils*. One of her less well-known novels was *En Herrgårdssägen* (1895), translated into English under the title *From a Swedish Homestead*.[29] Any person who bought this book expecting a sober, realistic account of Scandinavian rural life would be dismayed by its weird, dramatic plot, however. The novel's hero is the scapegrace young student Gunnar Hede, who neglects his studies to play the violin. To recoup his fortunes, he buys a large herd of goats and shepherds them through the cold Scandinavian landscape toward a market town. They are caught in a blizzard, however, and every single animal expires. After this tragedy, Hede loses his reason and becomes the Billy-Goat, an insane tramp who curtsies politely every time he meets an animal. One day, the Billy-Goat strays into a graveyard and sits down near a newly dug grave. As he is playing his violin, he hears a movement inside the coffin that has been put in the grave, and decides to unscrew the lid. Perceiving that the young woman inside the coffin is alive, he lifts her out. She turns out to be a girl named Ingrid, who was presumed to have died of grief after being badly treated by her wicked stepmother. She reflects on her narrow escape from a premature burial with these words: "What would have become of her, if he had not come. She would have woken up inside the dark coffin. She would have knocked on its lid, and cried out loud, but who would have heard her when she was buried alive under six feet of earth?" The Billy-Goat carries her back to the stepmother's house in a large bag. There the stepmother is busy telling her servant that they are just as well off without Ingrid and that her death was no great loss. When Ingrid steps out of the bag, the stepmother faints and the servant flees in horror. Ingrid is taken away by the Billy-Goat, and they have many further dramatic adventures before the ending of this remarkable novel. Selma Lagerlöf was well read in Swedish folklore, and it is likely that *From a Swedish Homestead* was inspired by two traditional folktales: one about the prematurely buried lady saved from the tomb, and the other about the love of a woman saving the reason of a man half transformed into an animal.

There is no evidence that Selma Lagerlöf herself feared premature burial, but a near-contemporary, the famous Danish storyteller Hans Christian Andersen, lived in fear of awakening in a coffin. Just like Wilkie Collins, he distrusted foreign doctors, and he always carried a card with him saying, "I am not really dead," which he put on the dressing table whenever he stayed at a hotel abroad, to prevent some careless doctor from wrongly declaring him dead. Two days before his death in 1875, Andersen asked a friend to make sure that his arteries were cut before he was buried.[30] A Danish contemporary of his, the eccentric writer Niels Nielsen, was even more frightened of premature burial. He suggested that every recently dead person should be put to bed, with a meal of cakes, wine, and beer ready on a table next to the bed, so that a tasty meal would be provided in the case of an unexpected resuscitation. After the doors and windows of the house of the deceased were shuttered, this building should be abandoned and a new one be built next to it for the family of the deceased![31] In Sweden, Alfred Nobel was another victim of the fear of premature burial. He ended the same will that stipulated the creation of the Nobel Foundation with these words: "Finally, it is my express wish that my arteries be opened after death, and that after this has been done, and competent doctors certified that clear signs of death are present, my body be burnt in a crematorium oven."[32]

SOME MINOR LATE nineteenth-century authors also used the premature-burial theme. The eccentric Englishman Frederick Rolfe, who called himself Baron Corvo, claimed that he had himself been buried alive by mistake when studying for the priesthood in Rome. Falling into a trance-like state after being frightened by a little lizard, he was coffined and walled up in a niche in an ancient church. He regained his strength, however, broke out of the coffin, and climbed down from the niche. This tale was published in a leading periodical and illustrated with drawings of "Baron Corvo's fearful experience" done under his own supervision.[33] Similarly ludicrous was Mrs. Harriet Lewis's sensation novel *The Haunted Husband*, which featured, among many other improbable occurrences, a premature burial and the substitution of a wax effigy.[34] In Gertrude Atherton's short story "The Dead and the Countess," a superstitious village priest is convinced that the dead in his churchyard are disturbed by the noise from the newly built railway line next to it. Every time the train comes by, he sprinkles all the graves with holy water and imagines that the dead people speak to him. One evening, he visits a local nobleman, the Count de Croisac, whose wife had been buried the day before. He tells the count that the dead people in his churchyard are being annoyed by the trains, and the nobleman believes him to be a madman. He then speaks of the frightful, smothered moaning sound he had just heard from the countess's grave, and the count rushes off, just in time to save his prematurely buried wife.[35] In Arthur Conan Doyle's short story "The Disappearance of Lady Frances Carfax," two Australian con artists kidnap a wealthy noblewoman and plan to have her buried alive. But Sherlock Holmes is hot on their trail. He intercepts the hearse, opens the coffin, and has Dr. Watson resuscitate the chloroformed Lady Frances.[36]

In a letter to his friend and fellow horror writer Clark Ashton Smith, H. P. Lovecraft discussed the problem of making a tale about premature burial plausible for the twentieth-century reader. In reference to one of Smith's stories, Lovecraft wrote that since the practice of embalming became universally accepted, the writer had to think up a reason for the individual's not being embalmed, before he could introduce this horrific theme in a story.[37] Several of Lovecraft's contemporaries did so. In John Dickson Carr's *The Three Coffins*, published in 1935, three Hungarian brothers imprisoned for bank robbery think of a daring method of escape, perhaps after reading Dumas's *Monte Cristo*. With the aid of a prison doctor, they are certified dead during a plague epidemic and buried alive in the prison cemetery outside the walls. The coffins are nailed shut, but the men have nail cutters with them in the coffins. One of them succeeds in breaking open the coffin and digs his way to the surface, callously leaving his brothers behind. One of them is later found suffocated in his coffin, the other bloody and insensible, but still alive.[38]

The American Cornell Woolrich added further horrors in his short story "Grave for the Living." In the beginning of the story, an insensitive police official shows the exhumed body of a war invalid to his wife and ten-year-old son: he demonstrates the broken fingernails, clawed coffin lid, blood-spattered shroud, and face contorted with horror, and tells them that the man was buried alive by mistake and slowly suffocated to death in his coffin. The wife faints on the spot and afterward goes insane; the son grows obsessed with the risk of premature burial. In his adult years, he becomes engaged to a wealthy heiress with the same morbid interest. He also becomes involved with a sinister secret society, called the Friends of Death, consisting mainly of people with an insane fear of premature burial. They claim to possess the secret of eternal life, and when one of their number died, skilled scientists administered a secret treatment, before the body was buried in a coffin with an air tube and an alarm. When returning to life, he would push a button to sound the alarm, and the other Friends of Death would come and dig him up. After such an experience, he should no longer fear either death or burial. In exchange, this ghoulish society demands steady payments of money and implicit loyalty from its members. The young man is prepared to give neither, and the Friends of Death exact their revenge. He is overpowered, shrouded, and put in a coffin to be buried alive, but tells them about his wealthy fiancée. They then release him, kidnap her, and bury her alive in his stead. With the help of the police, he finds her in time, and she awakes with "a shriek of unutterable horror" when the coffin lid is removed. The conclusion of this wild tale is that the police arrest the Friends of Death. It turns out that far from having discovered the secret of immortality, they had deliberately poisoned their own members with a narcotic drug and buried them alive; when rescued from the security coffins, they no longer feared death, and told other prospective members about their immortality. Woolrich himself feared premature burial and turned to the theme in at least one other story. He probably got the idea for "Grave for the Living" from Poe's tales, or possibly from one of the brochures about nineteenth-century security coffins.[39]

The prolific English thriller writer Dennis Wheatley made another use of the premature burial theme in his novel *The Ka of Gifford Hillary*.[40] A wealthy baronet is killed by a mad Welsh scientist, who has previously seduced his wife. The demented Welshman shoots Sir Gifford with a death ray he has invented, but the baronet is not quite dead. Just as the late nineteenth-century spiritualists had feared, his spirit has left his body, but has to return to it after burial. Fortunately, Sir Gifford had been fearful of being buried alive and equipped his coffin with air holes. In Michael Crichton's novel *The Great Train Robbery*, the security coffin known as Bateson's Belfry is put to an unexpected use.[41] The corpse of a young man is transported on a passenger train, when the bell on top of the coffin suddenly starts ringing! The lid is quickly unscrewed, and his sister throws herself down to embrace what she thought was her still living brother. "But she halted in mid-gesture, which was perfectly understandable. With the raising of the lid, a most hideous, fetid, foul stench rolled forth in a near palpable wave, and its source was not hard to determine. . . ." The girl faints, and the stench is such that the entire station platform becomes deserted. The coffin is quickly resealed and put into a compartment next to a safe where a lot of gold bullion was transported. But inside the coffin is a thief, who had smeared himself with the innards of a dead dog to produce a putrid smell; he is let out of the coffin by a colleague and proceeds to plunder the safe.

THE PREMATURE-BURIAL theme has also spawned a number of films, many of them loosely based on Poe's "The Premature Burial." The first was *The Crime of Dr. Crespi* from 1935, directed by John H. Auer and starring Erich von Stroheim as a wicked hospital doctor who plans revenge on a rival by plotting to have him buried alive. One of the episodes in the popular American 1960s TV series *Thriller*, hosted by Boris Karloff, was the 1961 "The Premature Burial," which also acknowledged a debt to Poe's tale. A wealthy man suffers from a cataleptic disease and becomes obsessed with the risk of being buried alive during one of his seizures. His unfaithful wife deliberately ensures that he is entombed alive, however. Karloff himself costars as a doctor who is a friend of the prematurely buried man and who eventually uncovers the truth about the wife's sinister doings. In 1962, Roger Corman's *The Premature Burial* appeared in the cinemas. The plot actually bears some resemblance to Poe's original story, in that the hero, Guy Carrell, played by Ray Milland, is a neurotic creature, obsessed with the risk of being buried alive after one of his cataleptic fits. He even has a security crypt constructed to make sure he will survive if the worst happens. Trying to prove to himself that he has been cured of his phobia, he opens his father's vault, but instead suffers one of his fits and is afterward found in a senseless condition. He is put in a coffin and lowered into a grave, and although a close relation of his has noted that he is still alive, she allows the funeral to go on. It is instructive that the transition between an accidental and a deliberate premature burial has been made in almost every cinematic adaptation of the theme. By this time, the fear of being mistakenly buried alive had become dormant, but the theme of a revival in the coffin still had its ghoulish fascination.

A still from Roger Corman's film *The Premature Burial.*

If Corman's film somewhat resembles Poe's original story, the opposite is true for *Edgar Allan Poe's Buried Alive,* directed by Gerard Kikoine and released in 1990. The heroine arrives at Ravenscroft Hall, a school for troubled teenage girls, to become the assistant to the charming and sinister psychiatrist who is in charge of this establishment. The school used to be an asylum for criminal lunatics, and the heroine soon begins to suffer nightmares about the brutal murders committed by one of these madmen. The obvious sequel is that the pupils start disappearing and that it turns out that the killer is still alive. Other directors have followed suit in discarding the theme of Poe's gloomy, morbid story and replacing it with a story line more familiar to the American adolescent filmgoer: that of a number of frightened, scantily dressed young blondes being sadistically murdered. Another strange "adaptation" of Poe's story, *Haunting Fear,* by the director Fred Olen Ray, also released in 1990, is eulogized on an Internet site by an enthusiastic amateur critic who recommends it to devotees of *Nightmare on Elm Street III* and other "splatter" B movies. Still available on video is also the 1991 movie *Buried Alive*, directed by Frank Darabont and starring Tim Matheson as a man poisoned and buried alive by his faithless wife and her doctor friend. This film was successful enough to spawn a sequel, *Buried Alive II* of 1997, directed by Matheson himself, and using much the same story line, except that it is now Matheson's niece, played by Ally Sheedy, who is deliberately buried alive.

More recommendable for the reader who wants to see a really thrilling film with the premature-burial theme, and also much more critically acclaimed than any of those mentioned above, is the Dutch film *Spoorlos*, directed by George Sluizer and based on a short story by Tim Krabbé. It was re-released as *The Vanishing* in 1988, with English subtitles. A young married Dutch couple take a break at a rest stop after returning from a bicycling vacation in France. The wife goes in to buy some cold drinks, and the husband never sees her again. He then grows increasingly obsessed with finding her, or at least knowing what became of her. For three years, he travels the country looking for her, puts advertisements in the newspapers, and even appears on TV with an appeal for witnesses. The uncertainty of not knowing what happened to her is driving him insane. Finally, he is contacted by the kidnapper, a clever psychopath, who offers to satisfy his curiosity, but only if the husband is willing to undergo the same experience as his wife. The murderer puts a sedative in the husband's coffee and asks him to drink—then he will wake up feeling exactly what his wife felt. The reader should be able to guess exactly in what situation he woke up. Not unexpectedly, this brutal, disquieting ending was a little hard to swallow for the American film moguls. George Sluizer was hired to direct a remake of the film with an American cast, featuring Kiefer Sutherland and Sandra Bullock as the married couple and Jeff Bridges as the murderer, and with a very artificial happy ending to the story.

XI

Were People Really Buried Alive?

A horror of great darkness all around
Lies on my soul; darkness that one may feel;
It interpenetrates, through all doth steel
Like a dead, icy, heart-chill; so profound
And numb, that very Being seems ice-bound!
O God! In mercy now reveal;
Break off this living tomb the sevenfold seal;
Buried alive, I feel as underground,
Cast me not off! From this dumb, living tomb.

—*H. Ellison,* Eclipse

As recently as a hundred years ago, several writers claimed that every year a very considerable number of people were buried before they were dead. A frequently repeated argument in alarmist propaganda was that the number of people falsely diagnosed as dead, but saved by lucky accidents, implied that many more victims of similar mistakes had been condemned to a hideous death below ground. None was more vehement than the aforementioned John Snart in condemning the careless and callous physicians who buried their mistakes alive: he claimed that for every apparently dead person saved from premature burial, there were one thousand others less fortunate. In Herr Köppen's previously quoted treatise, a German military doctor, connected with the Prussian general staff, stated that in his opinion one-third of humanity was buried alive. François Thiérry, *docteur-regent* of the Medical Faculty of Paris, claimed that one-third, or perhaps half, of all who died in their beds were not actually dead when they were buried. Johann Peter Frank was content with stating, from what evidence is not known, that he believed that the number of people buried alive was greater than the number of suicides.

In the nineteenth century, there was also much speculation about what proportion of human beings were buried before they were dead, with some sensational claims from the alarmists. Snart speculated that 10 percent of his countrymen were prematurely buried.[1] A Swedish physician offered the same rash conclusion: in his country, 1 out of 10 people were buried before they were dead.[2] Hyacinthe Le Guern thought 1 out of 2,000 buried were still alive when put in the earth.[3] His fellow propagandist Dr. Léonce Lenormand upped this figure to 1 out of 1,000 buried.[4] Writing in 1897, Dr. de Pietra Santa, editor of the *Journal d'Hygiène*, adjusted it downward to 2 out of 10,000.[5] At about the same time, the French alarmist B. Gaubert claimed that 8,000 Frenchmen were annually buried alive: Was the nation rich enough to survive this severe draining of its resources?[6] Using similarly dubious and unclear arguments, the Englishman J. C. Ouseley thought that 2,700 people per annum were buried alive in England and Wales,[7] but Dr. Stenson Hooker estimated that the annual number of prematurely interred Britons was closer to 800.[8] The Italian cleric Camillo de Lellis claimed that many hundred Italians were annually buried alive.[9] The American attorney Marvin Dana, writing in 1897, declared that an American was buried alive each week,[10] and Carl Sextus, another American, was convinced that 2 percent of his countrymen were prematurely buried. Sextus claimed to have collected 1,500 cases of premature burial over a period of eighteen years.[11] When the Fort Randall cemetery, near Rochester, New York, was removed in 1896, a Mr. T. M. Montgomery found that 2 out of 100 corpses bore evidence of having been buried alive. One of them was a soldier who had been struck by lightning; when the lid of the coffin was lifted, it was seen that the arms and legs had been drawn up as far as the confines of the coffin permitted. The other case was more ludicrous: the body of an alcoholic was found with the legs drawn up, and the hands clutching the clothing; in the coffin was also found a large, empty whisky flask. Mr. Montgomery concluded that this exhumation showed that approximately 2 percent of all Americans were buried before they were dead, just as Sextus had predicted.[12] Tebb and Vollum quoted this figure with approval, concluding that premature burial was definitely not rare. They claimed that their own 161 cases, and the even greater quantity collected by Franz Hartmann, certainly spoke for themselves.

A THOROUGH REVIEW of the literature on apparent death and premature burial would rather support the opposing view of Bouchut, Walsh, and other skeptics, that these figures were highly exaggerated. Many of the case reports reproduced by Hartmann and by Tebb and Vollum were brief extracts from newspapers, often told and retold by several papers before reaching the United States. In particular, the "From Our Foreign Correspondent" type of story, in which the hero is declared dead in the afternoon, is buried alive in the evening, and tells the story of his rescue to the reporters while eating a hearty breakfast the day after, is to be treated with great skepticism. It is well known that the more unscrupulous nineteenth-century reporters deliberately falsified thrilling stories to sell to newspapers, and accounts of premature burial were of high popular interest. In the 1860s, Eugène Bouchut, Alexander van Hasselt, and other skeptics were able to show many of these newspaper stories to be complete inventions and to demonstrate that even serious medical treatises on apparent death and premature burial were dependent on them. Nor is it possible to give much credence to cases of alleged premature burial where the individual has survived several days in an airtight coffin under six feet of earth. The eighteenth-century German physiologist Ernst Hebenstreit had made inquiries into the question how long a human being would survive under these conditions; he calculated that the air content could not last more than sixty minutes, and the majority of later writers on this subject have agreed.[13] In 1859, a Dr. von Röser wanted to settle this question by means of experiment: he buried a number of rats and mice alive in a coffin and dug it up after a couple of days to see how many were alive. The animals had kept alive by eating each other, however, and some had gnawed through the coffin and dug their way out. Dr. von Röser then decided upon another experimental strategy. He shut a large dog into an airtight coffin with a glass lid. After two hours, the struggles of the wretched animal intensified; the dog fell over and, after three hours, was declared dead. Dr. von Röser concluded that since the dog's body took up less room in the coffin than a human body, his findings were well in line with Hebenstreit's calculations.[14]

Much of the core evidence in the argumentation of the anti-premature-burial alarmists is derived from accounts of corpses and skeletons found in unnatural positions upon exhumation of coffins. Often, the newspaper accounts just said that it was "only too apparent" that the exhumed wretch must have been buried alive. Of the more detailed accounts, many are wholly unconvincing. Striving for sensational and horrific headlines, the writers made much of a cadaver found, for example, lying on its side; this could just as well be consistent with the coffin being tilted while put into the grave. If the corpse's face was contorted into a horrible grimace, or the arms and legs drawn up, there was ample scope for the imaginative journalist's talents to describe the horrid scuffle in the coffin, but neither of these changes is inconsistent with the natural composition of the body after death. It is well known that after death the corpse's musculature becomes relaxed, accounting for the serene and "smiling" countenance that has comforted so many of those recently bereaved, and given them the illusion that their dear departed had died at peace with the world. After this stage, cadaveric rigidity sets in, and there is a gradual contortion of the musculature, which may produce a hideous grimace on the face of the cadaver. Trained morticians, well aware of this phenomenon, sometimes have to use various covert operations, like cutting tendons and performing silicone injections, to maintain the feeble illusion of the beautiful death. The contraction of the muscles of the arms and legs, giving the false impression that the corpse had awakened and tried to push up the coffin lid, is also a perfectly natural phenomenon. The buildup of putrefactive gases inside a corpse is sometimes very considerable, and there are several well-described cases of well-seasoned corpses actually exploding, like that of King William the Conqueror, or bursting the coffin walls. The phenomenon of a shattered coffin in a vault, as in the Boger case, which I quoted initially to prepare the reader for the horrors lurking inside this book, is thus likely to have a perfectly natural explanation. In a much publicized case from Kronstadt, the skeleton of a boy, buried fourteen years previously, was found lying on the floor beside its coffin. The other coffins in the family vault were also empty and plundered. The premature-burial fanatics had a field day, imagining the desperate boy searching the coffins of his relatives for food, but the truth must be that grave robbers broke into the vault, which belonged to a well-known and wealthy family, in search of valuables.[15] Other cases of presumed premature burial may have been the result of the interference of body snatchers, who were after fresh bodies for dissection and who dumped the putrefactive ones back into the coffins without ceremony, or thieves of linen out to steal the shrouds. But I cannot disabuse the reader of all the fantastic fears engendered by the previous chapters. Not even the debunking forensic scientists of the present time have been able to explain a phenomenon reported in a few older instances of suspected premature burial: in addition to the contorted face, and drawn-up arms and legs, the clenched fists of the cadaver have been full of human hair, pulled out with the roots and matching that on the corpse's head.

A recurring theme in many eighteenth- and nineteenth-century reports of alleged premature burial was that the prematurely dead individuals were stated to have gnawed or eaten their own fingers, or even their entire arms. Modern forensic medicine has demonstrated that this was, most probably, due to rodents feeding on the dead body. They usually began at the fingertips and worked their way up the arms toward the torso. In 1966, there arose considerable uproar in the Austrian village of Vorarlberg, after it was believed that a groan had been heard from the coffin of an eighty-six-year-old, one-legged invalid who had died a few days earlier. The man's son insisted that a doctor be called, to open the coffin and examine the body. The discovery that the body had injuries on the face and fingers provoked an outcry that he had been buried alive and had bitten his hands in agony. The body was sent to the Forensic Institute of Innsbruck, where Professor F. J. Holzer examined it closely. He found that rats had gnawed a hole at the foot end of the coffin and had fed on the corpse's face, knee, and fingers. There was no vital reaction, proving that the man had really been dead at the time the rodents made a meal of him. Professor Holzer visited the old man's house, and although the son denied that they had rats, he found two large rat holes in the very room where the coffin had been kept. The rats had thus entered the coffin before it was carried to the mortuary. The groaning noise was presumed to have been either one of the rats squeaking or a mewing from one of the cemetery cats.[16] It is well substantiated from other modern observations that rats can cause injuries exactly matching those presumed to have been due to the frantic scuffle of the prematurely buried person, who gnawed his fingers and wounded his face in the desperate struggle to get free. This would almost put the mind at rest, were it not for an observation by the nineteenth-century Swedish physician Per Hedenius. Some young students had decided to walk across the frozen Öresund from Lund to Copenhagen; on their way back, however, the ice was broken by a storm. All the students were saved except one, who was found dead on a block of ice, where he had desperately tried to keep alive in the bitter cold. All his fingers were missing, and the doctor thought they had been eaten by some bird of prey attacking the body, but at autopsy their gnawed remains were found *in his stomach*.[17]

But what about the cases where a "cry for help" was heard from the coffin. Under the heading "Noises from the Tomb," Tebb and Vollum reported three of these instances. The *Sunday Times* of December 30, 1838, told the tale of a deceased Frenchman from Tonneins, who had been duly declared dead. But when the gravedigger threw the first shovelfuls of earth on the coffin, "an indistinct noise" was heard to emerge from it. Terrified beyond description, the gravedigger fled to seek assistance. The coffin was opened, and the man's frightfully contracted countenance was considered sure proof he had been buried alive. A doctor opened a vein, but no blood flowed; "the sufferer was beyond the reach of art." There was also a case of "smothered cries" coming from a coffin put in the cemetery watchhouse in a village near Naples, and another of "moaning" emerging from the coffin of a recently buried Russian girl. But when the coffins were opened by the trembling hands of the distraught vergers, who expected to see a hideous spectacle inside, the corpses were found to be dead and still.[18] These stories recall the screaming corpses of Garmann's miracles of the dead, and Michael Ranft's noisy cadavers that ate their own shrouds. They share a common origin in the same eerie phenomenon. In some instances, the buildup of intestinal gas inside the cadaver emerges through the throat, passing through the voice box and producing the so-called *Totenlaut*: a sometimes quite loud, moaning noise.[19] It is interesting to reconsider the famous case of the prematurely buried Madam Blunden, of Basingstoke, whose moaning in her coffin was heard by the playing schoolboys. The contemporary sources state that she was very stout and that the corpse was beginning to smell foully when it was buried. Interestingly, the late eighteenth-century popular tradition speaks only about a noise or moan coming from the Blunden vault; the words "Take me out of my grave!" may well be an invention by the author of the pamphlet *News from Basing-Stoak*, who probably wanted to give his readers a proper fright. Confirmatory proof comes from the description of Madam Blunden's body puffing up like a bladder when the coffin was opened; this definitely supports the conclusion that putrefaction was well advanced and the buildup of intestinal gases considerable. Perhaps, with this novel information at hand, the city fathers of Basingstoke should appeal the fines imposed against them in 1660.

But what about the horrific Swedish case that I also quoted in the introduction to curdle the reader's blood? The prematurely buried girl awoke in her coffin, crying out for help, and later gave birth to a child in her dark prison, before perishing in unimaginable tortures. But the groans and moans from the coffin were most probably due to the *Totenlaut*, just as in the cases discussed above. The buildup of putrefactive gases inside the cadaver would also result in a highly increased intra-abdominal pressure, which in some instances is strong enough to expel an unborn child from the womb. There were in fact quite a few such instances in the alarmist literature: both Hartmann and Tebb and Vollum provide several, and the newspapers delighted in reproducing (and sometimes inventing) these horrible anecdotes. In 1901, a pregnant Frenchwoman named Mme Bobin died of yellow fever in a hospital in Pauillac, in southwest France, and was duly buried. A nurse suspected that the burial had been overly hasty, and Mme Bobin's father had the body exhumed. It was found that a child had been born in the coffin, and in the resulting outcry the prefect and health officers of the town were fined the equivalent of £8,000 for their neglect. This was probably just as unfair as the Basingstoke sentence 240 years earlier.[20] The *Sarggeburt*, or childbirth in a coffin, was actually a well-described phenomenon in the forensic literature well before that time. In 1854, Alexander van Hasselt was called in to investigate a case of alleged premature burial, in which a woman had given birth to a child in her coffin. He was able to demonstrate that the child had been expelled postmortem.[21] A mid-nineteenth-century German article described a muffled explosion emanating from the coffin of a pregnant woman who had died twenty-four hours earlier. Upon investigation, it was obvious that the unborn child had been expelled from the womb with considerable force.[22] There were other, similar observations on record, and if Franz Hartmann or Tebb and Vollum had wanted to examine the truth of the newspaper anecdotes they quoted, the facts would not have been difficult to find. A German review from 1941 lists one hundred instances of childbirth in coffin and remarks that in the eighteenth and nineteenth centuries, many of them were originally reported as premature burials.[23]

FOR NATURAL REASONS, it has not been possible critically to reexamine the classical cases of purported premature burial. The one exception comes from Denmark, where the so-called mystery of Giertrud Birgitte Bodenhoff caused much stir in the 1950s. This immensely wealthy nineteen-year-old widow had died in 1798, after an unspecified illness lasting six days. She was buried in the family vault in the Assistens cemetery in Copenhagen. There was a persistent tradition, repeated in several Danish families, and particularly well known in Copenhagen, that something horrible had happened to the young widow *after* her burial. Two grave robbers had broken into the Bodenhoff family vault and opened the recently buried Giertrud Birgitte Bodenhoff's coffin to steal her jewelry. As they brutally tore one of her earrings off, the abscess in her ear, which had caused her "death" ruptured, and the woman awoke from her trance. The two robbers were startled to see her sit up in her coffin and look about her in a confused manner, successively taking in the full horror of her situation. The robbers were tough, determined villains, however, and although the young woman pleaded with them, and offered them money and a free passage to America if they let her live, they beat her to death with their shovels and threw her limp body back into the coffin.

Dr. Viggo Starcke, a Danish writer and physician, was a distant relation of the Bodenhoff family. When very young, he had heard his mother tell the story of how the apparently dead woman had been murdered in the vault, and this tale had naturally made a strong impression on him at this tender age. Much later, in 1952, he decided to find out the truth behind it.[24] According to the tradition, one of the grave robbers had confessed to a priest just before his own death many years after the horrid deed; the names of both the priest and the robber were known. The latter was Christian Meusing, head gravedigger at the Assistens cemetery in Copenhagen. Dr. Starcke found out, from the cemetery records, that Meusing lost his job in 1804 after it was found out that he had been involved with a gang of thieves who stole recently buried coffins as firewood. With this corroborative evidence at hand, Dr. Starcke applied to have the Bodenhoff family vault opened and Giertrud Birgitte Bodenhoff's coffin examined. If she had been brutally beaten to death, as the legend told, there should be fractures to show this; it was also likely that a severe abscess of the ear would have caused irreversible skeletal changes.

When the vault door was opened, the coffins could be seen to be in good order; when the lid of Giertrud Birgitte Bodenhoff's coffin was taken off, her skeleton appeared undisturbed. No jewelry was found, and Dr. Starcke of course presumed that the grave robbers had taken it; but there is no record that the woman had been buried with any ring or jewelry, except from an old, unverifiable tradition. Medical expertise could find no trace of any fractures, nor of any disease of the ear. There were signs of dental decay of considerable severity; this was rather fancifully presumed to be the cause of death, although the original sources mention nothing about any severe toothache during Giertrud Birgitte Bodenhoff's last days above ground. Much was also made of the skeleton's posture, and both Dr. Starcke and another medical man thought that the fact that the left foot was pulled up under the right lower leg would indicate that she had moved after being put into the coffin.[24] The discussion about the natural history of putrefaction inside a coffin quoted above would cast a good deal of doubt on this conclusion, however.

Although Dr. Starcke ended his book on the mystery of Giertrud Birgitte Bodenhoff by accepting the old tradition, but with the change that the robbers had strangled, not beaten her to death, he had little factual evidence to support his conclusion. Furthermore, he himself had discovered that her corpse had been kept above ground no fewer than five days before burial: Is it really possible that they would have buried her if there had not been signs of putrefaction after such a long period of time? Another difficulty was that the grave robber Meusing, who was presumed to have gone to America after the murder, had actually lived peacefully in Copenhagen until his own days were ended. Even members of the Bodenhoff family considered the entire tradition about the premature burial and murder of their kinswoman completely fabulous.[25] Another Danish historian later criticized Dr. Starcke's conclusions severely, particularly his unsubstantiated acceptance of Meusing as the culprit.[26] The strongest argument of all against the tradition about the premature burial and murder of Giertrud Birgitte Bodenhoff is of course, as the reader of this book knows, that it is a version of the extremely powerful and widespread legend of the Lady with the Ring.[27]

AFTER HAVING THUS effectively disposed of many of the old horrors of the premature-burial canon, we can turn to another crux in the arguments of Hufeland and other nineteenth-century writers: the *Scheintod*, or death trance. This obscure condition was described as a perfect counterfeit of death that convinced even experienced doctors that the patient had really expired. The body was ice-cold and still, the pulse imperceptible, and the respiratory movements no longer discernible. At the same time, however, the victim could still hear and feel everything that went on in the room, although she was incapable of moving or speaking. Modern medicine does not admit that there is such a disease, and historians have used this as an argument to doubt, or even to ridicule, Hufeland and his fellow believers in the *Scheintod*. But as we will see, it is not always valid to draw conclusions about the diseases of olden times using only arguments from present-day medical textbooks.

An instructive case of death trance was reported by the Austrian physician C. Pfendler.[28] In 1820, he had occasion to observe in Vienna a fifteen-year-old girl who had suffered from what was interpreted as intermittent epileptic seizures for not less than three weeks. The seizures were lengthy and severe, and Johann Peter Frank, who was consulted by Dr. Pfendler and his colleague Dr. Schäfer, feared for the worst. The following evening, the patient collapsed. The three doctors tried to revive her for several hours, by means of strong smelling salts and ammonia. They pinched her mercilessly and plunged needles into her feet, but the patient remained senseless and still. They next resorted to galvanism, but the patient showed no reaction. Even Frank was then prepared to admit defeat, but being well aware of Hufeland's teachings about the *Scheintod*, he advised that she be kept in a warm bed until putrefaction was apparent. After twenty-eight hours, the girl's relatives thought they could smell putrefaction from her corpse, and they wanted to prepare her for her funeral. The funeral bell was rung, the girl's lifeless body was dressed in a white gown, and her friends put a wreath of flowers in her hair. Dr. Pfendler was yet another disciple of Hufeland, however, and he came barging into the room, saying that he wanted to undress the corpse once more, to obtain definite proof that putrefaction had taken place. Such behavior from a medical attendant would today have been thought eccentric and rude, but the German people had become used to the excessive safeguards against premature burial, and the girl's relatives meekly walked out and let the doctor tear her clothes off. But Dr. Pfendler found no signs of corruption; instead, he thought he could see a weak respiratory movement of the chest. After having applied liberal amounts of a strong itching powder, Dr. Pfendler managed to induce the girl to open her eyes. With a smile, she said, "I am too young to die!" She later recovered totally and told her physician that although she had been completely unable to speak, move, or even open her eyes, she had heard and understood everything her doctors said. Dr. Pfendler did not believe her, but she convinced him by repeating some Latin words used by Frank during the consultation. The most horrific part was when she heard and understood that she was made ready for burial, without being able to make it known that she was still alive.

There were several other similar cases of death trance in the nineteenth century. In one instance, reported by Dr. M. Rosenthal of Vienna, a twenty-four-year-old woman was declared dead by one doctor, and another doctor advised her family to await the first signs of putrefaction before burial. A third, more competent medical attendant could feel no pulsations in the radial artery, and the pupils of the eyes did not contract when a torch was shone into them. There was no respiratory movement of the chest, but the doctor could observe a very slight movement of the sides of the abdomen. After clearing the room of people and shutting the door, he could hear a very weak heartbeat. Vigorous resuscitation was begun, and the woman recovered after forty-four hours of *Scheintod*. She had heard what had occurred during the later period of her strange disease and could repeat snippets of conversation spoken in the sickroom.[29] The Irish physician Thomas More Madden described the case of a young woman who remained in a death trance for five days: her body was ice-cold, the pulse was imperceptible, and the respiratory movements could no longer be detected. Electric stimulation elicited no reaction, but strong mustard plasters produced vesication. Laypersons thought her dead, but Dr. Madden, familiar with the literature on lethargy and morbid somnolence in younger females, advised continued observation over her lifeless frame. On the morning of the sixth day of her death trance, she suddenly awoke, called for her clothes, and wanted to come downstairs.[30] The London physician Dr. C. H. Miles observed a similar death counterfeit, lasting six hours, in a twenty-two-year-old woman. Her temperature was subnormal, her pulse could not be felt, but upon auscultation with a stethoscope her heart sounds could be indistinctly made out. Dr. Miles feared that if the attending physician did not take care, a patient with this kind of death trance could actually be buried alive.[31]

The Philadelphia neurologist Silas Weir Mitchell was once consulted by the parents of a histrionic young woman, who was suffering from remarkable deathlike spells. These spells could be provoked by some person's teasing or annoying her, or indeed if certain topics were brought up in conversation; before the consultation, her parents solemnly warned Dr. Mitchell what topics were taboo. But Dr. Mitchell was unkind enough to talk about just one of these subjects, probably because his curiosity had got the better of his good manners, and he was keen to see one of her death trances firsthand. She immediately said, "I am going to have an attack; feel my pulse! In a few minutes I will be dead!" The pulse rate went up, and the patient fell back in her easy chair. She then said, with what Dr. Mitchell recognized as the typical indifference of the hysteric, "Now, watch it, you will be amazed!" Within a few minutes her pulse rate fell—indeed, completely disappeared—and the heartbeat became inaudible. She grew deathly pale, still, and cold, with no perceptible signs of life, and remained in this condition for two days, except for short periods when her heart beat very quickly, as it did when she eventually recovered, without the slightest ill effects. It was clear to Dr. Mitchell that this was a disease of the mind rather than of the body.[32] When the famous French neurologist G. Gilles de la Tourette reexamined some of these remarkable nineteenth-century cases of death trance, he agreed that this was an uncommon hysterical disorder, which he termed lucid hysterical lethargy.[33] A review of the relevant nineteenth-century literature thus clearly shows that Hufeland had been right: there existed a hysterical disorder that matched his description of the *Scheintod*. Like other of the more extreme forms of hysterical conversion, the lucid hysterical lethargy become much more rare, if not altogether extinct, in the twentieth century. It has been suggested that cataplexy, a variant of narcolepsy causing abrupt and reversible paralysis, could mimic the old *Scheintod*, or apparent death, but the cataleptic attacks in this disorder last at most thirty minutes, and ocular motility and breathing are not disrupted.[34]

We have seen that a fair proportion of the old cases of presumed premature burial have a natural explanation; the risk of being buried alive was no doubt greatly exaggerated by the contemporary alarmist writers. So were there *any* real instances of people being prematurely buried? We know about a considerable number of cases, from the seventeenth century right up to our own time, of people who were falsely declared dead, but who revived before burial. Many of them were vouched for by competent physicians, and their genuineness is not in question. But these narrow escapes are not enough: the individual wanting to pooh-pooh the eighteenth- and nineteenth-century fear of premature burial could just claim that the vigilance of the doctor in charge, or the precautions taken to safeguard against a premature burial, had had their desired effect. The postmortem changes in the cadaver discussed in the previous section make, it is very difficult to draw any conclusions whether an individual was alive or dead when buried from the attitude of a skeleton or cadaver. Another conclusion from the previous chapters is that the frequent distortion, or even invention, of newspaper reports of premature burial deprives these records of any credibility. The same goes for the alarmist anti-premature-burial pamphlets of Lenormand, Le Guern, Hartmann, and others of that ilk. We need cases where a living individual was shrouded, coffined, and put into the earth, but later dug up and proved to be alive by an unbiased, competent physician. These are not easy to find. There is a phenomenon in modern science called publication bias, meaning that journals and editors are much more likely to publish positive than negative data. For example, results from a successful trial of a new drug or procedure are much more likely to get published than results from a similar, but unsuccessful one. Similarly, a case report gets published if the patient is cured, but is rejected if the patient dies, as a result of the treatment. The same forces operated in the nineteenth century, although not to the extremes of today. Premature burial was an inflamed topic, and particularly after Bouchut's discovery became generally accepted in the 1860s and 1870s, any physician who actively supported the anti-premature-burial activists was considered a traitor by his peers. It would take a brave author (and a brave editor) to publish a report of a case of a still-living person being declared dead by a physician. The strong professional loyalty made it highly compromising for any doctor to name and shame one of his colleagues for a fatal error of this kind. The doctor who signed the certificate was of course even less likely to publish the event in any way. Rather understandably, there are no case reports on record with titles like "I falsely diagnosed my patient as dead" or "My patient was buried alive."

However, in 1823, the newspaper *Schwäbischer Merkur* had a short note about a premature burial not far from Stuttgart. Dr. von Jäger, a local practitioner who was critical of the *Leichenhaus* movement, decided to investigate, hoping to prove that the case was a newspaper fiction.[35] He spoke to the people involved, including the doctor who had been called to help the alleged victim. A forty-year-old shoemaker had been declared dead after a long malady on February 9, 1822, and was buried three days later. His family testified that he had looked dead, but there had not been much putrefactive stench and no cadaveric rigidity at all. Because of either professional loyalty or other concerns, Dr. von Jäger treaded lightly on the subject of exactly who (a local dignitary or a medical practitioner?) had allowed the funeral to go ahead. The man was buried in a six-foot-deep grave. When the gravedigger had almost filled the entire hole with earth, he could hear a hollow sound of knocking coming from the head end of the coffin. Two other men also heard it, and they frenetically started digging to save the man. The soil had been stamped down hard, and it took them almost half an hour before the coffin was broken open. The gravedigger, seeing the man inside move, screamed, "Er lebt!"—"He lives!" The man lay in an unnatural position, with the arms drawn up, and when a doctor arrived on the scene, he had to admit that he certainly did not look like a three-day-old corpse. The man was not cold, and there were no livid spots. When a vein was opened, blood flowed freely and stained the shroud red. Vigorous attempts at resuscitation were begun once the man had been transported to the nearest house: the soles of his feet were cut with a knife, and sal ammoniac spirit blown into his nose. But these efforts were fruitless, and late on February 13 the man was buried a second time. Dr. von Jäger argued that the corpse might have been turned over in the coffin when a very shaky ox-drawn carriage transported it to the churchyard, and that there were no marks on the hands and feet. But he could not explain the knocking from the coffin, nor why the blood had still flowed from the corpse. Even more sinisterly, the doctor in charge had freely admitted that after the man had been declared dead a second time, his corpse developed rigidity and livid spots exactly like a *one-day-old cadaver*. Medical practitioners of this time had considerable knowledge about the postmortem changes in cadavers, and this information cannot be interpreted in any other way than that the man had been buried alive.

Another verified instance of premature burial comes from France. In 1867, when cholera was raging in the French departement of Morbihan,

A curious French death certificate, reproduced by Dr. Séverin Icard in *La Presse Médicale* of August 17, 1904, telling that Antoinette Rouzeyrol, spinster, aged twenty-one years, was declared dead at three o'clock in the morning of March 23, 1902. The additional writing underneath adds that she was *declared to have returned to life* at ten o'clock in the morning the same day.

the twenty-four-year-old Philomèle Jonetre suddenly fell ill, with generalized weakness and a severe headache as the main symptoms. The doctor who was called suspected cholera. As Mlle Jonetre's condition quickly deteriorated, the local priest came to administer the last sacraments. In the evening, she appeared to have died. The body was put in a coffin one hour after her presumed death and was buried in a grave six feet deep five hours later. It is again not mentioned whether she was examined by a doctor after death and who (the priest?) allowed this hasty funeral. But when the earth was poured onto the coffin, Philomèle Jonetre knocked against the coffin lid, and this was fortunately heard by the gravedigger. The coffin was speedily disinterred and the woman brought out alive. A certain Dr. Roger examined her and also spoke to the people involved; he had no doubt that she had been buried alive. Philomèle Jonetre died a few days later and was buried a second time. The case was published in 1870, not by Dr. Roger, but by the forensic specialist Alphonse Devergie.[36] In 1874, the French skeptic M. Tourdes, an associate of Dr. Bouchut's who investigated reports of alleged premature burials, wrote to Dr. Roger to inquire whether he was sure the patient had really been alive when taken from the coffin. The doctor rather unwillingly replied that this had definitely been the case: she had breathed and there had been distinct rhythmical sounds in the region of her heart. There was no muscular rigidity, and Dr. Roger clearly saw one of her muscles contract and her eyelids move. The influential forensic scientists Paul Brouardel and Georg Puppe accepted this case as one of live burial.[37] Dr. Puppe presented another one. A German gravedigger was carrying the coffin of a fourteen-day-old child to the grave. The child had not been seen by a doctor either before or after death, it was claimed, but this had not prevented the authorities involved from screwing the coffin shut and sending it on its way to the grave. But when the coffin was put in the grave, the gravedigger heard a weak whimper coming from within and opened it. To his horror, the child was alive, and moved its arms and legs. It died the next day.[38]

In the picture archives of the Wellcome Institute in London are two curious photographs, alleged to depict a Chinese beggar from the city of Nanking, who had been saved from a premature burial.[39] They are undated, but probably date back to the early twentieth century at least. According to the inscription, the man was presumed to be dead from paralysis and was duly buried in what must have been a shallow grave on the hillside. More than a week later, people heard knocking coming from inside the coffin, and the police and cemetery authorities were informed. On opening the coffin, they found the man alive; though emaciated in the extreme, he was taken to a hospital and eventually recovered. Without the photographs, this story would have appeared very unlikely indeed, particularly since it was stated (as in many invented stories of premature burial) that the man's paralysis had been completely cured by his lengthy stay in the narrow house underground.

It is very unlikely that the cases detailed above are the only ones:

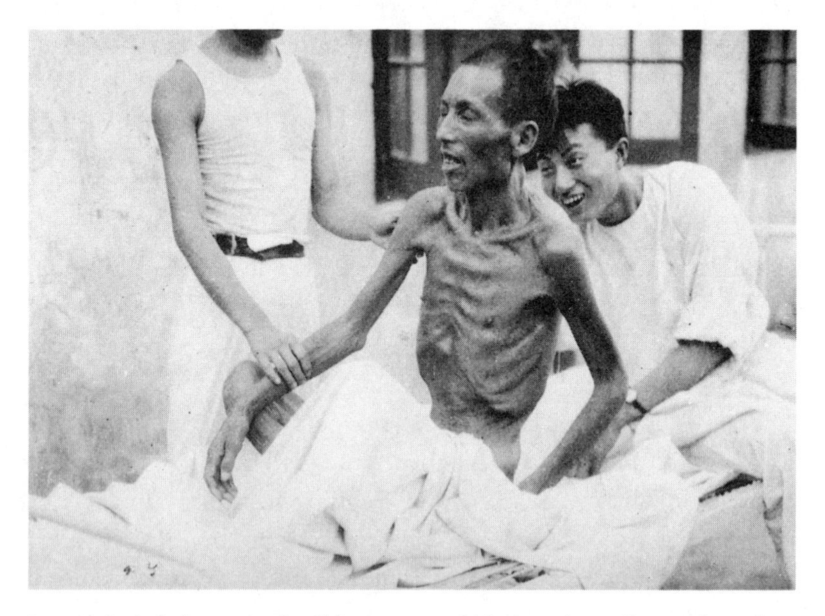

An undated photograph of a Chinese man said to have been the survivor of a premature burial. *Copyright the Wellcome Trust.*

from the time when coffins and burial vaults first came into use, it must have occurred regularly, but probably not frequently, that living people were buried by mistake. The safeguards instituted by Bruhier and his followers probably reduced the number of premature burials, as did the gradual acceptance, in the nineteenth century, of the notion that a doctor should be called to examine the dead. Still, a few genuine cases happened even in the nineteenth century, particularly during cholera epidemics. It was the practice to bury the dead hastily during these epidemics, in order to try to stop the contagion from being passed on, and the shortage of doctors made it impossible for them to examine every deceased person. Furthermore, patients in the so-called cold stage of cholera may become surprisingly deathlike, without losing their chances of eventually surviving the infection. It is no coincidence that the famous painting *Inhumation précipitée*, by the Belgian artist Antoine Wiertz, portrays a cholera victim coming to life in his coffin. The sheer number of people dying each year, all over the world, and the sometimes uncertain methods of declaring people dead in developing countries, indeed imply that, even today, people may be at risk of being prematurely buried.

XII

Are People Still Being Buried Alive?

> There was a young man at Nunhead,
> Who awoke in a coffin of lead;
> "It is cosy enough,"
> He remarked in a huff,
> "But I wasn't aware I was dead."
>
> —*Anonymous limerick reproduced in* Burial Reformer *1 (1908): 100*

After the Great War, the popular fear of premature burial decreased considerably. Although both medical and nonmedical writings on the subject continued to appear, the London Society for the Prevention of Premature Burial and similar organizations elsewhere declined. This was due largely to the medical breakthroughs of the time, which led to an increased confidence in the medical profession. But it does not mean that "accidents" like those recounted earlier were totally eradicated. It still occasionally happened that some presumed corpse came to life in the morgue, or even in the coffin, but the subject had lost some of its power to fascinate the public.

In the summer of 1915, Dr. D. K. Briggs of Blackville, South Carolina, was called to attend the thirty-year-old black woman Essie Dunbar, who had suffered an attack of epilepsy. He found no signs of life and declared her dead. The corpse was put in a wooden coffin and the funeral arranged for eleven in the following morning, to give Essie's sister, who lived in a neighboring town, the chance to participate. Although the ceremony was a lengthy one, with three preachers taking turns to perform, the sister had still not arrived when Essie's coffin was lowered into its six-foot-deep grave. She appeared a few minutes later, however, and the ministers agreed to dig up the coffin so that she might see Essie one last time. But when the screws were removed and the coffin lid opened, Essie sat up in her coffin and smiled at her sister. The three ministers fell backward into the grave, the shortest suffering three broken ribs as the other two trampled him in their desperate efforts to get out. The mourners, including Essie's sister, believed that she was a ghost, and fled yelling. When they saw that Essie, who had climbed up from the grave, was actually pursuing them, they stampeded into town in a state of complete hysteria. For many years, Essie Dunbar was viewed with suspicion in the neighborhood; there were rumors that she was a zombie who had returned from the dead. In later life, she became a popular local personality, and it is by no means unlikely that the story of her resurrection from the tomb was somewhat improved upon as it was told and retold, and finally appeared in the newspapers after her second and final death in 1955.[1]

Probably the most remarkable twentieth-century instance of alleged premature burial, although dependent on newspaper evidence, is that of the Frenchman Angelo Hays, from the village of St. Quentin de Chalais.[2] In 1937, when he was nineteen years old, his motorcycle skidded out of control; he was thrown from the vehicle and hit a brick wall head first. When Dr. Bathias, the local practitioner, attended him, he shook his head: young Hays was a dreadful sight, with a serious head injury. There was no pulse or respiration, and the doctor could hear no heartbeat with the stethoscope. Angelo Hays was declared dead and his body taken to the local morgue. His parents were not allowed to see the corpse, since the injury looked too dreadful. Three days after the accident, his body was buried; the coffin was carried to the grave by eight of his friends from the local volunteer fire brigade. But in nearby Bordeaux, an insurance firm discovered that Angelo Hays's father had recently insured his son's life for 200,000 francs. An inspector was sent to St. Quentin de Chalais to investigate the accident. A farmer confessed that his tractor had leaked oil, and that this had caused Hays's motorbike to skid out of control, but the determined inspector decided to have the body exhumed to ascertain the exact cause of death. Two days after the funeral, Angelo Hays's coffin was dug up and taken to a forensic institute in Bordeaux. When the doctor in charge removed the shroud, he felt that the body was still warm! Hays was taken to a hospital in an ambulance; after several operations and a long period of rehabilitation, he recovered completely. He had been deeply unconscious during his two days in a coffin underground. The earth had been very dry, and had not been stamped down over the coffin; it was supposed that he would have suffocated, had the head injury not led to a diminished demand for oxygen.

After his resurrection, Angelo Hays became quite a celebrity in France. For many years, people traveled from all parts of the country just to see and to speak with him. In the 1970s, Hays decided to try to make money from the fame that followed his amazing rescue from the tomb. He invented a security coffin that was something of an improvement on Count de Karnice-Karnicki's late nineteenth-century model. This supercoffin cost £4,500—as much as an automobile—and lacked nothing in luxury, being equipped with thick upholstering and a soft pillow for the head. It was high enough for the buried individual to be able to sit up, and had a library of books the victim could read while awaiting rescue. The coffin's food locker was supplied with rations like those of the American astronauts. The controls at the coffin's dashboard maneuvered the oxygen supply from large gas tubes, the ventilation fan, the air pumps, the chemical toilet, the electrical alarm, and the shortwave radio transmitter and receiver, which worked through an aerial sticking up above the earth. Several extra gadgets, like a small oven, a refrigerator, and a hi-fi cassette player attached to the radio, were optional. Hays went on tour with his coffin and became a media star. During the shows, he let himself be buried in the coffin, to demonstrate that it was fully functional. At a demonstration in Bordeaux, 25,000 paying spectators came to see him, and the take from this and a series of other shows brought him a handsome profit, although few, if any, of his security coffins found buyers. Hays himself boasted that a wealthy, ninety-three-year-old Frenchwoman had ordered two coffins, one for herself and one for her niece; the niece presumably did not much appreciate this macabre present from her dismal *tante d'heritance*. At one of the shows, French television buried a camera and a microphone in the coffin with him, and the sprightly Angelo Hays sang his favorite songs to a live TV audience; although the acoustics in the narrow coffin cannot have benefited his performance, the program was a great success. The Monte Carlo television company signed a contract with him, and he recorded a series of thirteen shows with them during a five-year period. No one has doubted the amazing story of Angelo Hays, and the fact that no journalist exposed him at the height of his fame may indicate that he was really buried alive in 1937. The only inconsistency in the various accounts of his story is that one version holds that his body was exhumed owing to insurance technicalities; another, that his coffin was disinterred so that his uncle could see his body for the last time. A third version, from a notoriously unreliable source, maintains that he suffered a heart attack and actually awoke in his premature grave, before being saved after a visitor to the churchyard heard his frantic banging against the coffin lid.[3]

Angelo Hays's security coffin was not the last one on the market. In the 1960s, the body of a man named Archibald Maclean was exhumed in Detroit, since there was a suspicion that he might have been poisoned. The corpse was found in a horrible, unnatural position, and there was much newspaper speculation that he had been buried alive. According to a German writer, the consequence of this was that more than three thousand Americans decided to make sure they were not buried alive, by requesting that their bodies be cremated or embalmed, or deliberately mutilated to safeguard against a live burial. An American firm of funeral directors could also supply a security coffin, fitted with an alarm and a seventy-two-hour supply of oxygen. The multimillionaire John Dackeney had a huge security vault built for him at a churchyard in Tucson, Arizona, fitted with an alarm and with automatic steel doors that opened for three hours every night for the first twelve weeks. After his death in 1969, hundreds of curious onlookers gathered every night to see if Mr. Dackeney would walk out when the steel doors opened, but always in vain.[4] In the mid-1990s, the mayor of Apareidan, Brazil, built a security vault with four airholes for ventilation.[5] At about the same time, a newspaper reported that the Tuscan watchmaker Fabrizio Caselli marketed a coffin equipped with a bleeping device, a telephone, a flashlight, an oxygen tube, and a heart stimulator. It sold for £3,000, and Caselli told the press that if sales were brisk, he hoped to establish three medical centers in Italy, whose sole purpose would be to respond to premature burial emergency calls.[6]

TEXTBOOKS ON THE current principles for declaring people dead advise caution in cases of head trauma (like that of Angelo Hays) and epilepsy (like that of Essie Dunbar), as well as in drowning, lightning stroke, and electrocution. Hypothermia also requires special care: a person found "frozen to death" should not be declared dead until he or she is both warm and dead. In particular, hypothermia in combination with drug poisoning with narcotic substances can produce a remarkably deathlike state, without affecting the individual's chances of recovery. Knowledge of this phenomenon had existed since the mid-nineteenth century; for example, one of Dr. Josat's patients with no audible heartbeat had taken an overdose of some narcotic substance. It was further highlighted just after the Great War. In October 1919, the twenty-three-year-old nurse Minna Braun went into a Berlin pharmacy to purchase some morphine and veronal. She then proceeded to a park in the freezing cold weather and tried to commit suicide by swallowing a very large dose of these drugs. She quickly lost consciousness and showed no signs of life when found the next day. At the local mortuary, the doctor on duty could detect no evidence of breathing or heartbeat, could provoke no reflexes, and described the body as deathly pale and rigid. Hot sealing wax was dripped on her skin as a test of life, but it elicited no reaction. Minna Braun was declared dead and her body put in a coffin. A detective came to the morgue fourteen hours later to identify the deceased. When the coffin was opened, the mortuary assistant was aghast to see that the "corpse" moved its head. The same doctor who had declared her dead came rushing in, and this time he heard a slow heartbeat with his stethoscope. Braun was put in a hot bath and vigorously brushed with a strong prickly brush; from time to time, artificial ventilation had to be resorted to. Wisely, the doctors decided to apply the stomach pump, and considerable amounts of morphine and barbiturates that remained in the stomach were flushed out. The patient was then put in a hospital bed. Slowly but steadily, Minna Braun recovered and was able to walk out of the hospital; one hopes that her narrow brush with death deterred her from any further suicide attempts. The doctors in charge described the case as a medical mystery: How could a human being have existed without circulation or respiration for at least seventeen hours?[7]

We today know that the requirement for oxygen decreases with lower body temperature, given that the natural defense mechanisms against hypothermia, like shivering, are nonfunctional, something that can be accomplished by an intoxication with barbiturates or other narcotic substances. At a body temperature of 20 degrees Celsius, the body's need for oxygen is just 15 percent of the normal oxygen intake, and this can be further lowered by a concomitant intoxication with barbiturates or other drugs with a depressing effect on the central nervous system (CNS). In such an extreme state, there may be just ten (or even fewer) heartbeats per minute, and just two to three respirations. It is impossible to feel a pulse or to detect any spontaneous respiration. Electrocardiography (ECG), which has been used to distinguish between life and death in difficult cases at least since the early 1930s,[8] may be fallible in these extreme cases, since the ECG complexes seen are so severely deformed as to resemble artifacts. Similarly, electroencephalography (EEG) may be fallible, since the depression of the CNS may be profound enough to result in an isoelectric EEG, with no sign whatsoever of spontaneous brain activity.[9] Quite a few people every year try to commit suicide by taking an overdose of barbiturates or other CNS depressants while outdoors. Most of them die as a consequence of the overdose; a few are discovered while still conscious and are thus saved; and a small proportion are brought into hospital in a deathlike state of extreme hypothermia, facilitated and deepened by the drug overdose. Most of these survivors are lucky enough to encounter an experienced emergency physician and to be taken into intensive care. Some others may be examined by a young, inexperienced doctor in a busy emergency room at a small hospital. This doctor is probably more concerned with keeping the conscious patients in the emergency ward alive than with spending a long time examining the lifeless cadaver of what appears to be yet another suicide. It is quite possible that under these conditions the patient can be wrongly certified to be dead. This happened to a thirty-six-year-old woman in Sweden in 1969[10] and a twenty-three-year-old woman in Liverpool in 1970.[11] Recognized to be alive by paramedical personnel prior to removal to the mortuary, both recovered without ill effects. In 1970, the French journal *La Presse Médicale* detailed three cases of people who had attempted suicide by taking high doses of barbiturates. Despite careful monitoring of their life signs, they had been erroneously diagnosed as dead; one of them had actually recovered in his coffin.[12]

After the early 1970s, these remarkable modern cases of apparent death appear to have become a "damned" phenomenon, which the medical establishment no longer wishes to acknowledge; the later instances have been reported in the newspaper press only. Some of these newspaper reports are reasonably well verified, containing the names and sometimes photographs of the individual falsely declared dead; it also increases the credibility if the case has been reported by several independent national newspapers. In 1986, a twenty-seven-year-old drug abuser was found lifeless in woods near Reigate, in Surrey, not far from London. According to an article in the *Daily Mail*, he was certified dead at the New East Surrey Hospital, Redhill. In the morgue, a technician heard spluttering sounds from the alleged corpse, but attempts at resuscitation were in vain: the man died thirty-six hours later.[13] In 1988, a man in Mons, Belgium, was diagnosed as being both heart- and brain-dead after severe exposure to the cold; he revived after having long remained in a comatose state, with a very low heart rate.[14] According to another newspaper account, the forty-year-old Brooklyn schoolteacher Nancy Vitale was declared dead by Emergency Medical Service technicians after being found in her basement apartment after a drug overdose in June 1993. There was no pulse and no recordable blood pressure, and the body was lifeless and rigid. Two hours later, she gave a spluttering sound when the medical examination for determination of the cause of death was to begin; two days later, she was awake and could speak.[15] In 1995, sixty-two-year-old Daphne Banks was certified dead by her doctor after a drug overdose. Mrs. Banks was taken to the mortuary of Hinchingbrooke Hospital, Huntingdon, United Kingdom, but was later heard snoring by a family friend who had come to take a final farewell of her. After being rushed to the hospital intensive-care unit, she recovered fully.[16] Fortunately, there are today quite a few safeguards to prevent the premature burial of one of these unfortunate individuals—particularly the examination by the police to determine the suicide's identity, and the examination by mortuary workers when the corpse is prepared for burial. But if the patient's stomach is not pumped, more CNS depressant will be absorbed as the body warms up, which will slow the awakening from the coma. In these conditions, since the body's need of oxygen is much less than usual, it would have the effect of prolonging survival inside the coffin, and a confused awakening in the narrow house below ground is a distinct possibility.

The reader of this book must be interested to know whether I myself have encountered, during my fifteen years as a practicing physician, any instance where the diagnosis of death was unclear. In 1989, I had occasion to observe a woman who had been taken to the emergency ward in a comatose state, with an extremely low body temperature. She was an elderly alcoholic who had, on New Year's Eve, celebrated the occasion with several bottles of vodka, seated on her balcony in freezing cold weather. She fell asleep out there and was well nigh dead when finally brought in; her body temperature was far below what was considered, at the time, to be consistent with human life. The body was cold and deathlike, and there was no pulse to be felt; the policemen considered her to have expired, and her skinhead nephew mourned her as dead. The electrocardiogram revealed a heart rate of twenty and waning. Through gradual extracorporeal warming of the blood, by means of a heart-lung machine, her life was saved.[17] This treatment was remarkably successful, and the woman recovered without any ill effects. Her behavior was more than a little odd, however, and the consultant anesthesiologist and I decided to speak to her relatives, raising the question whether she had suffered some residual brain damage. They replied, "No, she has always been like that!" Her marked obesity, isolating the inner organs from the cold, was probably to be thanked for the fortunate outcome of this case. I told her, in no uncertain terms, that she had been extremely lucky to have cheated the Grim Reaper, hoping that this experience would wean her from the bottle and prevent her unnecessary death from alcoholic cirrhosis of the liver. But unlike the Lady with the Ring and the other resuscitated heroines of eighteenth-century fiction, she did not go on to have twenty-six children, weave prodigious amounts of linen, and praise the Lord in diligent prayer. All I could get from her was the comment that this experience had, indeed, taught her never to drink heavily—while seated outside in subfreezing temperatures!

IN 1930, THERE was a brief scare in France about the risk of premature burial, with an appeal to the Chamber of Deputies and some medical writing on the subject.[18] In 1943, a German forensic specialist published an article on newspaper stories about apparent death and premature burial, proving that, even in the middle of a devastating war, the German people's fascination with the *Scheintod* continued unabated. He contacted various colleagues at the hospitals where apparently dead people were supposed to have risen from their coffins or awakened in the morgue. The majority of the newspaper stories turned out to be complete fabrications. One of these untrue stories allegedly took place in Istanbul: a crippled old farmer was declared dead by a doctor, but revived in his coffin the next day. He then walked into the hospital and struck the careless doctor on the head with one of his crutches; the old invalid's blow was powerful enough to crack the skull of the doctor, who fell dead to the floor.[19]

From 1945 until 1965, there was less writing about apparent death and the risk of being buried alive than in any period after 1740. But the specter of premature burial did return with a vengeance. In 1979, the first alarmist book on the subject in more than fifty years appeared, written by Jean-Yves Péron-Autret, an obscure French medical practitioner.[20] A friend of his, Professor Louis-Claude Vincent, had once been employed to check the water supply for the great American war cemeteries in Holland, Belgium, Luxembourg, and France. Observing that a good deal of excavation work was going on in the cemetery, Vincent inquired about its purpose. He was told that this was a top-secret operation to determine how many of the soldiers had been buried alive and recovered consciousness inside their coffins. It concluded that 4 percent of the American soldiers showed unmistakable signs of having been buried alive. All 150,000 graves were opened and the remains put into new mahogany coffins after being rearranged to remove traces of broken fingers, nails embedded in the coffin lid, and skeleton hands full of human hair.

Dr. Péron-Autret went on to claim that the proportion of American soldiers buried alive had been the same in the recent Vietnam War; in the United States, two hundred heart attack victims were buried alive every day. In florid prose, he wrote that the technological advances that had taken some of their countrymen to the moon was of no avail to save these wretched Americans from the appalling fate that their ancestors had feared most of all. Dr. Péron-Autret had spoken to sixty French gravediggers, all of whom told him that they had at least once dug up coffins containing evidence that living bodies had been buried in them. According to Professor Vincent, 10 percent of all French people were buried before they were dead. Dr. Péron-Autret went on to claim that in Spain forty-seven people had been buried alive between 1976 and 1978. One of these cases, concerning a Spanish woman alleged to have been saved from a premature burial after spending six days buried in a coffin, was described in grisly detail: "she had scraped so hard upon the lid of the coffin that her fingers had been reduced to a mutilated mash of blood and bone, whilst her elbows, from the constant pressure, had been flayed of all skin, so much that the actual tendons were visible. Could it be said that this wretched, bleeding lump of flesh was still a person? . . . still an individual? Or had it indeed been reduced to 'a thing'?"

Dr. Péron-Autret provides no evidence, no literature or newspaper references, even for his wildest claims, and his book is a worthy successor of those of the most fanatical nineteenth-century anti-premature-burial propagandists, like Hyacinthe Le Guern and Léonce Lenormand. These two had been relatively sincere and philanthropic individuals, but there is reason to suspect that Dr. Péron-Autret was more knave than fool. His friend Professor Vincent had invented something he called bio-electronics, a method to detect life in the tissues of human beings, and Dr. Péron-Autret demanded that this method be developed throughout the world. To describe the magnitude of the medical disaster happening all around the world, he claimed that Professor Vincent knew that 50 percent of all coronary victims were only apparently dead, and that 300,000 Americans were thus buried alive every year, for want of "bio-electronics."

Dr. Péron-Autret's book was translated into English in 1983, but it failed to raise the same stir abroad that it had created in its country of origin. It was mentioned in some alarmist German articles, and a similar volume soon appeared in German. While Dr. Péron-Autret's deplorable volume is likely to have sprung from impure motives, the journalist Claus Boetzkes's book *Scheintot begraben* was prompted by a series of German instances of people being erroneously declared dead by qualified doctors.[21] One example was the eighty-year-old Emma Sikorski, who had been found lifeless in her freezing cold house. The body was rigid, there were no discernable pulsations, and no heartbeat could be heard. Her pupils did not react to light. The doctor signed a death certificate stating the cause of death to have been "cardiac apoplexy," and the old woman's body was taken to the undertaker's shop. Here she moved her hand and gave a sigh as the shroud was fitted. It turned out that, like many other victims of the twentieth-century variety of the *Scheintod*, she had taken an overdose of barbiturates. The reason her pupils had not reacted was that she had previously undergone cataract operations in both eyes. The careless doctor gave back his fee, but Frau Sikorski's daughter successfully took him to court, and he had to pay 14,000 marks in damages. This was not the only near-fatal mistake in declaring people dead in Germany at this time. After five much publicized instances in 1976 and 1977, the forensic specialist Professor Hans Mallach spoke up in the media, declaring that the German medical profession was losing the confidence of the public; physicians should always rule out that the patient had taken a drug overdose before declaring him or her dead. Professor Mallach claimed that one patient, the forty-one-year-old Monika Abelein, might have been buried alive had a policeman not seen her move while transporting her to the morgue in a van. There was much renewed interest in the *Scheintod* and its victims in Germany at this time, and the production of sensationalist newspaper and magazine articles almost resembled that in the 1790s.[22] The medical debate on the subject also gathered new strength, some writers deploring that an old superstition was again raising its ugly head, and demonstrating that the alarmist articles were sometimes inventing or re-using cases in the time-honored manner described earlier. Others followed Mallach in demanding burial reform, suggesting that a corps of specially trained doctors for the dead, whose only task was to verify that people were really dead before burial, be established throughout Germany.[23]

IT IS DIFFICULT not to link this sudden outbreak of a renewed fear of apparent death and premature burial, in both Germany and France, to the introduction of a new definition of death. Whole-brain death, as defined by the Harvard Criteria in 1968, means a permanent loss of all brain function; persons in such a state were defined as being dead, and their organs could be transplanted into others. This redefinition of death was still being relatively smoothly introduced throughout the Western world during the 1970s and 1980s. For a person with a lingering fear of premature burial, the image of the doctor whetting his scalpel near the unconscious body of an accident victim, ready to harvest the organs from this "dead" person whose heart was still beating, was not a pleasant one. A deplorable feature in a 1967 issue of *Newsweek* was entitled "When Are You Really Dead?" and described the transplant surgeons as ghouls ready to rip the poor accident victim open for the sake of his organs.[24] With characteristic bluntness, Dr. Péron-Autret boldly opposed the concept of brain death, claiming that many people declared brain dead could still be nursed back to life. He demanded that the ignorant doctors no longer be allowed to "reign over a race of zombies, totally at their mercy when it comes to matters of life and death." Nor could Claus Boetzkes resist quoting some newspaper reports of people returning to life after being declared brain dead. At least one of these, concerning Michael McEldowney, who started to breathe during the operation to harvest his kidneys, appears to have been a true story.[25]

The certification of whole-brain death requires evidence of cerebral unresponsitivity, as determined by a neurological examination. In many countries, an EEG is mandatory; in some, also a carotid angiography or a test for brainstem-evoked potentials. Brainstem function must also be absent, as determined by a failure of spontaneous respiration and other neurological signs. The diagnosis for this extensive brain damage must be known, and the presence of hypothermia, drug intoxication, or other metabolic or physiologic disturbance must be ruled out with absolute certainty. In studies involving many hundreds of patients, it was found that a person fulfilling these criteria always suffered a cardiac standstill within two weeks.[26] Similarly, a prospective study of more than one thousand patients showed that individuals surviving severe head injuries would not have been suspected of being brain dead even in their worst state soon after injury.[27] I have been able to find only one case report in the medical literature that directly challenges the criteria of brain death, by presenting the case of a patient who satisfied these criteria, and had a flat-line EEG, but who still survived and recovered.[28] Competent experts demonstrated this report to be completely unconvincing, however, since there was no diagnosis of irreversible, structural brain damage and because the patient had taken two drugs acting as respiratory depressants.[29] In 1980, the BBC television documentary *Panorama* criticized the current criteria for declaring people brain dead, giving several American examples of tragic and sinister mistakes. The medical profession objected strongly to this sensationalist documentary, however, and could demonstrate that the cases presented, although mostly true, were all instances in which the criteria for brain death had been incompetently applied, rather than that the criteria were unreliable.[30] Of four patients, three had taken drug overdoses, and a fourth was alleged to have been "purple and thrashing about" when an American neurologist allegedly declared him brain dead—a most unlikely scenario, because spontaneous movement excludes this diagnosis. Nevertheless, the number of donor kidneys in Britain fell considerably in the months after the *Panorama* documentary. This shows both the unquestionable power of today's mass media and the irresponsible way this power is often used, in this instance most likely causing the unnecessary deaths of several renal failure patients.[31]

During the 1980s and 1990s, brain death became generally accepted. A Swedish study published in 1994 revealed that the fear of the brain-dead organ donor's not being really dead was a very minor factor in determining people's attitudes toward organ donation.[32] The fear of being buried alive has also become dormant, and although some scare-mongering books were published in the 1990s, they failed to impress the reading public.[33] In the United Kingdom, the well-defined concept of brainstem death—defining the brainstem as the critical system of the brain, controlling respiration, consciousness, motor output, and sensory input—has gained prominence. In the United States, the murkier concepts of "neocortical death" or "persistent vegetative state," involving loss of higher cerebral functions, have been much discussed. But whereas a person with a destroyed brainstem will always die within two weeks without regaining consciousness, individuals with neocortical brain damage can still breathe and maintain blood pressure and body temperature, although they may suffer irreversible loss of consciousness and cognition.[34] Prognostication is much more difficult with higher brain criteria: it is not just in cheap TV movies that people sometimes unexpectedly come out of lengthy comas. Doctors and philosophers alike have objected that the introduction of higher brain criteria would mark the first step down a very slippery slope indeed. One of their arguments is that the concept of laboratories where these still breathing "corpses" were maintained for regular organ harvesting, whenever the need occurred, would be repugnant to the average person and lead to a loss of confidence in the medical establishment.

THE OLD STORIES about reviving corpses, which had existed in European folklore long before Bruhier's books were published in the 1740s, still live in the tabloid newspapers, along with tales of alien abductions, crop circles, spontaneous human combustion, and snakes living as parasites in the human stomach.[35] In particular, the legend of the Careless Anatomist has considerable staying power in modern folklore. In one version, a corpse revives in a New York mortuary and grasps the Anatomist's throat; the doctor falls dead from the shock. In another version, the living corpse starts struggling with an undertaker before succumbing to a heart attack; in a third, two timorous doctors drop dead from shock when a "corpse" suddenly sits up and bursts into a drinking song.[36] Another of these old legends, that of the Lecherous Monk, is revived in a recent, distasteful newspaper story, adapted to suit the shallow morality of the reader of sensationalist newspapers. The corpse of a young girl revives after being raped in the mortuary by a perverted grave robber; it is usually added that the girl's parents did not press charges against the rapist, but instead thanked him for their daughter's life.[37] In one popular newspaper myth, the reviving "corpse" is considered a ghost by the family and relatives;[38] in another, the corpse sits up in the coffin after being struck by lightning.[39]

The readers of tabloid newspapers are easily amused, and most of the tales about reviving corpses are much less elaborate than these. The headline is adjusted to appeal to the vulgar mind and the low intellect: "*Dead* Wrong," "An Undertaker's *Moving* Story," or "A *Grave* Mistake." It is deemed extremely funny if the dead granny revives in the mortuary and all the guests run out in terror, or if the dead man leaps out of the coffin and runs away. The latter motif is repeated in many variations. Many of the reports are from Italy or Australia; for some reason, the journalists seem to credit the Mediterranean and antipodean peoples with astounding powers of resurrection.[40] Occasionally, the dead man or woman jumping out of the coffin meets an uncertain fate. In one story, the Maltese Joseph Cremano leaps out of the coffin and jitterbugs down the church aisle, before falling dead in front of the astounded congregation.[41] In another, an apparently dead Romanian woman recovers inside her coffin and opens the lid while it is actually lowered into the grave; she jumps out of the coffin and runs out into the road, where she is hit by a car and killed.[42] An amusing variation, known from Japan and Australia, involves a naked streaker arrested by the police; he is released when he proves that he has actually jumped out of a coffin in a morgue.[43]

Sometimes, an added twist brings in the age-old concept that the Grim Reaper must have his due, although the victim might be changed. When the Sicilian Antonio Percelli jumps out of his coffin alive and well, his mother has a heart attack and drops dead by the graveside.[44] In a story from Saudi Arabia, a man falls from a windmill and his family presume that he is dead and bury him. After a day, the prematurely buried man awakes to the sound of grazing sheep walking above his grave. His screams are heard by some shepherds, and he is dug up. With remarkable resilience for someone who has just survived a premature burial, he walks home in his burial shroud. When his mother and sister see him, they go mad and drop dead. According to one version of this story, "which could not be officially corroborated," the prematurely buried man goes mad as well and winds up in an asylum.[45] Any readers still inclined to believe what they read in the newspapers should compare two clippings from London newspapers, dated 1992 and 1995: "Tragic Julie Carson dropped dead in a New York funeral parlor when her mother Julia came back to life, sat up in her coffin and asked what was going on" and "Connie Palmer died in a funeral parlor in Wellington, New Zealand, after her heart-attack victim mum Carol sat up in her coffin."[46] The reworking of invented old news as new was a commonly used technique among Victorian journalists and remains current today.

An equally ludicrous newspaper myth is that of the bumpy ride in the hearse. When a wealthy American is taken to his funeral in a luxury hearse, a tire blows, and the hearse skids through the front window of a rival undertaker's parlor. The hearse's back doors fly open and the coffin is flung into the premises. The hearse driver and the mourners are astonished to see the alleged corpse walking out through the shattered glass, dressed in his white burial robes.[47] In a variation, a thief steals the hearse and crashes it into a tree; the rescue team that goes to the scene of the accident is astonished to hear a squeaky voice from the coffin plead, "I'm not dead! Don't bury me!"[48] In the full-blown urban legend on this theme, a student is hired as a temporary hearse driver. He drives along a road with many potholes, and the corpse in the coffin revives and knocks on the lid. The student stops at a phone booth and calls an ambulance, but is too impatient to await it and drives the hearse, at breakneck speed, toward the nearest hospital. Halfway there, he crashes into the ambulance and the woman in the coffin is killed.[49]

But not all newspaper stories of people mistakenly declared dead are frauds, myths, or hoaxes. As we know, quite a few recent reports of people reviving after taking drug overdoses are beyond reproach. There are also a few instances of people erroneously declared dead through the gross carelessness of the individual signing the death certificate. For example, eighty-six-year-old Mildred Clark was declared dead by a New York coroner (not a doctor) and put in a morgue cooler. She had been there for an hour and a half when the morgue supervisor saw her move.[50] The coroner was hounded by the press. In April 1996, the fifty-nine-year-old diabetic Maureen Jones collapsed in her cottage in Thwing, Humberside. She probably had eaten too little, which had caused the blood glucose to fall to dangerously low levels. Her son alarmed her family doctor, who came to see her patient not long afterward, but without finding any signs of life. Maureen Jones was declared dead, and a hearse came to collect her body. But when a sharp-eyed police constable saw her foot move, she was rushed off into intensive care. The careless doctor was sued, as she deserved to be. Maureen Jones was still alive in 1999.[51]

As we saw, the leaders of the anti-premature-burial movement of the 1890s were facing an uphill struggle; the political and medical establishment stood united to treat their alarmist propaganda with a mixture of disgust and ridicule. The doctors were particularly annoyed by the propagandists' blatant distrust of the medical profession. It was in those years that this form of aberrant behavior was actually *pathologized.* In 1891, the Italian psychiatrist E. Morselli described what he termed a novel form of mental aberration: taphophobia, or the exaggerated fear of being buried alive.[52] The patient he described had put a clause into his will stipulating that a candle, food, and drink be put in his coffin after he was presumed to be dead and that the coffin be equipped with airholes and an alarm. Despairing even of this safeguard, he finally desired to be stabbed in the heart as a more definite method of prevention. A German psychiatrist described a similar case not long thereafter: a hypochondriac, neurasthenic man lived in fear of premature burial and frequently visited his family doctor to be reassured that his arteries would be cut after death.[53] In 1942, an American psychiatrist reported two cases of taphophobia, which he described as an idea not infrequently encountered in psychoanalytic practice.[54] The obsessive fear of being buried alive usually began in childhood and became a frequent source of nightmares and bed-wetting. A man suffering from severe taphophobia indulged in fantasies of being buried alive day and night and insisted that his coffin be equipped with electric wiring and an air tube. A review on taphophobia published in 1977 adds two more cases: a fifty-two-year-old woman who had lived in fear of premature burial for forty years and her eighteen-year-old daughter, who shared the same predicament.[55] The daughter was the victim of a severe anxiety neurosis. No further articles on taphophobia have been added to the scholarly literature, but although this diagnosis is rarely made in modern European psychiatry, there are indications that some American phobia specialists are not unfamiliar with the condition.[56]

In spite of Morselli's claim, taphophobia was of course not a "new" diagnosis in 1891; what was new was only the declaration that this form of aberrant behavior was pathological and the coining of a name for the "disease." An extreme case of taphophobia had been described already in 1862, concerning an English lady who feared premature burial so much that it embittered her life.[57] Some historians have chosen to extend the definition of taphophobia to embrace all individuals who feared being buried prematurely, like Mme Necker, Arthur Schopenhauer, and H. C. Andersen,[58] but to do this is to ignore the definition of phobia as an *irrational* fear. If a person living in 2001 would have his or her life embittered by anxiety about being wrongly declared dead and prematurely buried, he or she would be considered to be suffering from a highly unusual and distressing syndrome of aberrant behavior. After all, every modern doctor would agree that the risk of being accidentally buried alive is much less than that of falling victim to one of the perils of modern industrialized state: being killed or maimed in a traffic accident, or shot in the street by a robber. The modern psychiatrist would probably suspect either that the person had had a distressing experience at an early age, involving an undead "corpse," or that the taphophobia was a kind of repressed fear of death. But the eighteenth- and early nineteenth-century people who were anxious about the uncertainty of the signs of death had support from a considerable part of the medical establishment, as well as several prominent public figures. Famous doctors and scientists like Winslow, Buffon, Frank, Hufeland, and Julia de Fontenelle agreed that the risk of a premature burial was considerable, and leading citizens like Duke Ferdinand of Brunswick, Mme Necker, and Cardinal Donnet advocated burial reform to stop this abuse. Thus the fears of these people in former times can by no means be considered *irrational*, but must be regarded as a product of their environment; the deep, primal fear of awakening in the coffin after burial was fueled by the medical and social establishment, and the pan-European epidemic of taphophobia that followed was no wonder.

It is known that the prevailing fears of society can be influenced by the contemporary medical debate; for example, bacteriophobia (the fear of contagion) did not exist before the discovery of microorganisms as transmitters of disease and has become scarcer after the discovery of effective antibiotics. In his original treatise, Morselli described his taphophobia patients along with some cases of another psychological aberration—dysmorphophobia, the delusion of being remarkably ugly. It is instructive that while taphophobia has become increasingly rare, an epidemic of dysmorphophobia is raging in the United States, whence it has spread to the wealthy, overconsuming European countries. Brought on by unrealistic body ideals and the power of mass suggestion provided by late twentieth-century advertisement techniques, and fueled by the plastic surgery industry's desire for further business, the rage for cosmetic surgery shows no sign of abating. As an epidemic of aberrant behavior, it is more extensive, and probably more deleterious, than the habit of worried Victorians to go shopping for coffins with an ejection seat.

AS WE KNOW, the mid-eighteenth-century process of dechristianization of death facilitated the spread of the fear of premature burial. But both the traditional fear of hell and the new fear of being buried alive were expressions of a deeper, underlying fear of death. A remnant of the child's natural feeling of omnipotence—to be immortal and the center of the world—remains in most people, and this irrational belief serves the beneficial purpose of protecting against the fear of death. But this brittle psychological defense fails in a child or adolescent frightened, at an impressionable age, by a thunderous sermon describing the torments of souls in hell, or terrorized by a sadistic pamphlet about scratched coffin lids, bloody shrouds, and convulsively clenched skeleton hands full of human hair. The fear of being buried alive has another parallel with the traditional belief in the afterlife: neither the wretch condemned to burn in hell nor the person who awakened in a coffin to scream and scratch the unyielding coffin walls could ever return and tell others about his experiences. This inherent uncertainty about exactly how many people were buried alive explains some of the subject's fascination. It was stressed by the anti-premature-burial campaigners, who boldly asserted that as many as one-tenth of humanity was buried alive. As

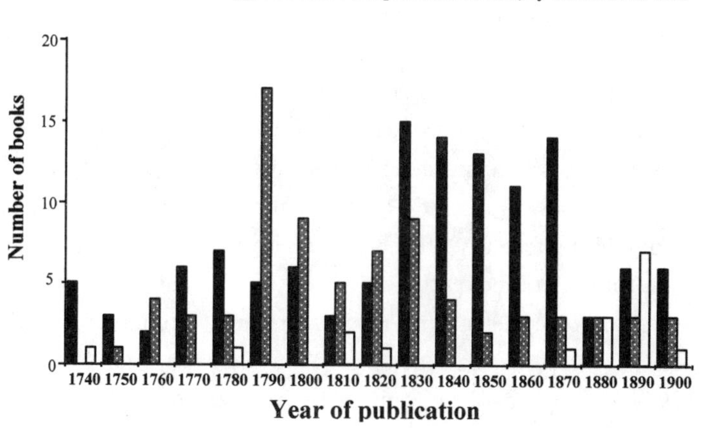

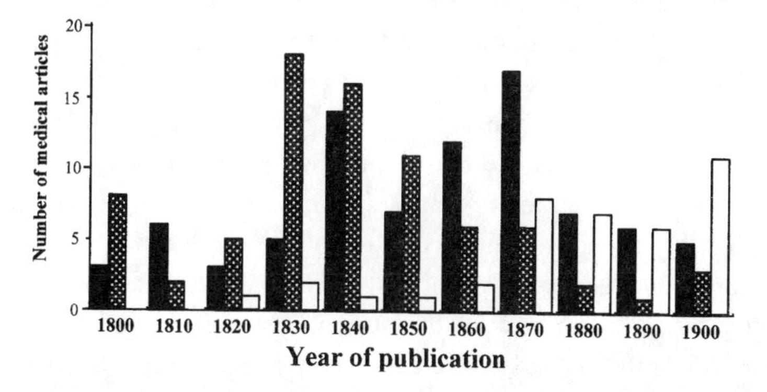

The number of books and pamphlets (*above*) and medical articles (*below*) on apparent death and the risk of premature burial plotted over time, in France (black bars), Germany (shaded bars), and Britain and the United States (white bars).

we noted, however, modern forensic science has revealed that the findings of corpses and skeletons in strange and macabre attitudes, which inspired so many horror stories about the horrid scuffle in the coffin, were actually not incompatible with putrefactive postmortem changes or with the wounds inflicted by rodents feeding on the dead body. It is impossible to draw conclusions, from exhumation findings alone, how many people were actually buried alive in the eighteenth and nineteenth centuries. There are numerous instances on record, from the seventeenth century right up to our own time, of people who were mistakenly diagnosed as dead by doctors or laypeople, but who afterward recovered in morgues or hearses, or even in their coffins. There are also a few accounts, confirmed by reliable medical professionals, of people actually buried alive who recovered in their coffins and called for assistance by knocking on the lid. These few reports are very unlikely to represent the total number of premature burials at this time. Given a strong bias against publication, many cases have probably been suppressed by concerns for the family of the prematurely buried person, or a wish to conceal the identity of the doctor who had made the fatal error.

So, are people still being buried alive? The sheer number of people dying each year, some of them in underdeveloped countries where the legal framework for the verification of death resembles that of nineteenth-century Europe, renders it difficult to answer this question in the negative. In the United States and the rest of the Western world, efficient safeguards against premature burial are in place, in particular an obligatory medical examination of the deceased.[59] But even these safeguards have been proven to be fallible in some instances of gross incompetence on the part of the attending physician and in several twentieth-century cases of severe hypothermia after intoxication with CNS depressants. The German nurse Minna Braun was very lucky not to be buried alive after a suicide attempt in 1919, and in the 1970s a young Frenchman was detected to be alive when the lid was about to be screwed onto his coffin in the undertaker's shop. Others may have been less fortunate. Indeed, the best preventive measure for those fearing a premature burial may well be to avoid taking a drug overdose while outdoors in cold weather. A few of these presumptive suicides are likely to have been buried alive, and it is by no means impossible that some of them actually awoke in their coffins after burial, to die a second death much more horrible than they had intended. Today, it would be possible to construct super-mortuaries, staffed by doctors specially trained to detect states of apparent death and equipped with modern technology to assess the biological activity of various organs. But the costs for such facilities would be immense, and modern health economists would hardly consider it cost-effective to build mortuaries all over the United States to save just one or two people a year.

If the medical profession's attitude toward apparent death and premature burial depends on the individual doctor's confidence in applying the currently used signs of death, the popular fear of premature burial is a function of the public confidence in the medical profession at large. That profession was relatively confident in the signs of death in the early 1700s, but its confidence dwindled after the publication of Bruhier's books; the most extreme expression of this decline was the establishment of the German waiting mortuaries. The fierce mid-nineteenth-century premature-burial debate in France resulted from a rise in the medical profession's confidence in the signs of death, at a time when the public confidence in doctors remained low, at least partly because the irresponsible newspaper press published many invented cases of people being buried alive. Not until the 1880s could Eugène Bouchut and his adherents declare themselves the victors of this debate, and not until the 1910s did the organized anti-premature-burial movement collapse. A minor premature-burial scare in continental Europe in the late 1970s and early 1980s was very likely a reaction to the redefinition of death to whole-brain criteria. But in spite of some horror-mongering books, and a spate of magazine and newspaper articles, this scare was of brief duration. The medical profession closed ranks around the new death criteria, and the spread of general education meant that people were less susceptible to the alarmist propaganda and more disposed to trust the medical establishment. With regard to the brain-death controversy, people were probably influenced by the concurrent horror propaganda about brain-dead people maintained on life-support; the fear of being kept alive indefinitely in a comatose state served as an effective antidote to any residual fear of premature burial. The growing practices of embalming and cremation also effectively deter fear of premature burial; American morticians were right to claim that any corpse that had passed through their embalming workshop would never wake up again. Another argument is that the distance between the dead and the living is greater in the late twentieth and early twenty-first centuries than ever before. Many Americans have never seen a corpse, and those who have probably encountered only the travesty of a beautiful death concocted in the mortician's beauty parlor. Nor is the prevailing present-day lifestyle in the United States and large parts of the Western world, set by egotistical, hyperactive people obsessed with amassing money and luxury goods, conducive to gloomy contemplation of death or fear of a possible life after burial. Nevertheless, I will conclude this book with the following thought, from John Weever's seventeenth-century book of epitaphs:

So many burials, reader, in one book!
Warn thee, that one day, thou for death must look.

NOTES

INTRODUCTION

1. Edgar Allan Poe's fascination with premature burial is more fully discussed in chapter 10, below.

2. J. Snart, *Thesaurus of Horror; or, The Charnel-House Explored* (London, 1817), pp. 84–87.

3. M. B. Lessing, *Skendöden eller det Osäkra i wår Kännedom on Lifwets Utslocknande* (Linköping, 1837), pp. 61–62, 125–26.

4. F. Hartmann, *Premature Burial* (London, 1896), pp. 2–4.

5. J.-J. Bruhier, *Dissertation sur l'incertitude des signes de la mort*, vols. 1–2 (Paris, 1746–49).

6. W. Tebb and E. P. Vollum, *Premature Burial and How It May Be Prevented* (London, 1905).

7. See *Burial Reformer* 1 (1905): 11, and 1 (1906): 35.

8. Lessing, *Skendöden*, p. 127.

9. P. Ariès, *L'homme devant la mort* (Paris, 1975), published in translation as *The Hour of Our Death* (New York, 1981).

10. J. McManners, *Death and the Enlightenment* (Oxford, 1981), pp. 48–49.

11. My *Cabinet of Medical Curiosities* was originally published by Cornell University Press in 1997; a W. W. Norton paperback appeared in 1999. There are also Swedish, Spanish, Portuguese, Japanese, and Korean language editions of this book.

12. "Buried Alive," produced by Dream Catcher Films for the Discovery Channel. My own contribution was filmed in the crypt underneath the Kensal Green cemetery in London, full of decaying coffins.

13. I wish to thank the British Library, the Library of the Royal Society of Medicine, the Wellcome Institute Library, and the Guildhall Library, all in London; also the National Library for the History of Medicine in Washington, D.C.; in Germany, the Staats- und Universitätsbibliothek in Göttingen and the Deutsche Bibliothek in Frankfurt am Main; in

Austria, the Österreichische Nationalbibliothek of Vienna. In Holland, the Universitietsbibliotheik Leiden, the Museum Boerhaave Library, and the Royal Library of The Hague were consulted; in Sweden, the Lund University Library and the Waller Library of Uppsala; in Denmark, the Royal Library of Copenhagen.

14. The German nation's fascination with *Scheintod* and premature burial continued well into the twentieth century, and not fewer than seven Ph.D. theses on this subject, and three popular books, have appeared since the 1960s. Among the theses, *Die Angst vor dem Scheintod in der 2. Hälfte des 18. Jahrhundert*, by M. Patak, Zürcher Medizingeschichtliche Abhandlungen, n.s. 44 (Zürich, 1967), *Scheintod und Todesangst*, by I. Stoessel (Cologne, 1983), *Der Scheintod, eine medizingeschichtliche Studie*, by E. Vogl (Munich, 1986), and *Das Leichenhaus*, by M. U. Stein (Marburg/Lahn, 1992) deserve particular mention. The popular books are lacking in academic credentials, although both the scare-mongering *Scheintod: Auf den Spuren alter Ängste*, by S. Schäfer (Berlin, 1994), and the whimsical *Lebendig Begraben*, by T. Koch (Augsburg, 1996), are quite well researched. In the French language, the most valuable book is the erudite *Mort apparente, mort imparfaite*, by C. Milanesi (Paris, 1991). There are no full-length academic works in English, but the article by M. S. Pernick in R. M. Zaner, ed., *Death: Beyond Whole-Brain Criteria* (Dordrecht, 1988), pp. 17–74, is particularly well researched. Other modern articles include those by A. de Bernardi and P. B. Bollone in *Giornale di Batteriologia e Virologia* 82 (1969): 915–46, by J. F. V. Deneke in *Herz/Kreislauf* 1 (1969): 435–41, 486–94, by M. Alexander in *Hastings Center Reports* 10 (1980): 25–31, by P. Vecchi in *Revue de Synthèse*, ser. 3, nos. 113–14 (1984): 143–60, by P. Mouchet in *Annales de la Société Belge d'Histoire des Hôpitaux et de la Santé Publique* 28 (1991): 35–49, by S. M. Quinlan in *French History* 9 (1995): 27–47, and by R. Olry in *Vesalius* 2 (1996): 111–17.

15. Schäfer, *Scheintod*, and C. E. Boetzkes, *Scheintot begraben* (Percha, 1984).

16. Boetzkes, *Scheintot begraben*, pp. 115–32, and Vogl: *Der Scheintod*, pp. 86–88.

17. *Transactions of the Royal Humane Society* 1 (1794): 441.

CHAPTER 1: MIRACLES OF THE DEAD

1. For an overview of death and burial in classical society, see J. M. C. Toynbee, *Death and Burial in the Roman World* (London, 1971), and I. Morris, *Death-Ritual and Social Structure in Classical Antiquity* (Cambridge, 1994). For reviews on the signs of death and their practical application during this period, see C. Milanesi, *Mort apparente, Mort imparfaite* (Paris, 1991), pp. 67–78, and the articles by L. G. Stevenson in *Bulletin of the History of Medicine* 49 (1975): 482–511, H. Grassl in *Grazer Beiträge* 12–13 (1985--86): 213–23, and M. S. Pernick in R. M. Zaner, ed., *Death: Beyond Whole-Brain Criteria* (Dordrecht, 1988), pp. 17–74.

2. C. Plinius, *Natural History*, ed. H. Rackham, Loeb Classical Library (London, 1942), 2:619–31.

3. C. Celsus, *De medicina*, ed. W. G. Spencer, Loeb Classical Library (London, 1935), 1:109–17.

4. The interesting work of the pseudo-Quintilien is discussed by D. R. Shackleton Bailey in *Historical Studies in the Physical Sciences* 88 (1984): 113–37, and Milanesi, *Mort apparente*, pp. 71–75.

5. The Greek novels are commented on by D. W. Amundsen in *Bulletin of the History*

of Medicine 48 (1974): 320–37); the Arabic sources, by J. C. Bürgel in *Zeitschrift für die Geschichte der Arabisch-Islamischen Wissenschaften* 4 (1987–88): 175–94.

6. These novels are commented on by Amundsen in *Bulletin of the History of Medicine* 48 (1974): 320–37, and Milanesi, *Mort apparente*, pp. 75–78.

7. General works on death in the Middle Ages are T. S. R. Boase, *Death in the Middle Ages* (London, 1972), P. J. Geary, *Living with the Dead in the Middle Ages* (Ithaca, 1994), and C. Daniell, Death and *Burial in Medieval England* (London, 1997).

8. M. Wagner, *Die Bedeutung des Scheintodes aus rechtsmedizinischer Sicht* (Munich, 1982), p. 17.

9. R. Wilkins, *The Fireside Book of Death* (London, 1990), p. 16.

10. See C. Gittings, *Death, Burial and the Individual in Early Modern England* (London, 1984), p. 30.

11. See the article by R. C. Finucane in J. Whaley, ed., *Mirrors of Mortality* (London, 1981), pp. 40–60.

12. Daniell, *Death and Burial in Medieval England*, pp. 116–24.

13. See V. Møller-Christensen, *Bogen om Æbelholt Kloster* (Copenhagen, 1954), pp. 252–54, and the article by S. M. Hirst in M. Carver, ed., *In Search of Cult* (Woodbridge, 1993), pp. 41–43.

14. Deliberate live burial was used as a punishment for a variety of crimes—sodomy and bestiality, in particular—in medieval Europe. See K. von Amira, *Die germanischen Todesstrafen, Abhandlungen der Bayerischen Akademie der Wissenschaften, philosophisch-philologische und historische Klasse*, vol. 31, no. 3 (Munich, 1922), 150–55, 213–35.

15. H. Kornmann, *De miraculis mortuorum* (Frankfurt, 1610); the section on Archbishop Geron is in chap. 60.

16. C. F. Garmann, *De miraculis mortuorum* (Leipzig, 1709); the section on resurrected corpses is on pp. 1200–44.

17. For a learned discussion of Garmann's remarkable book, see P. Ariès, *Essais sur l'histoire de la mort en occident* (Paris, 1975), pp. 123–31, and idem, *The Hour of Our Death* (New York, 1981), pp. 353–61.

18. M. Ranft, *Tractat von den Kauen und Schmatzen Der Todten in Gräbern* (Leipzig, 1734). This was the German edition of his *De masticatione mortuorum in tumulis* (Leipzig, 1728).

19. P. Forestus, *Observationum et curationum medicinalium* (Leiden, 1590), book 17, obs. 9.

20. T. Kirchmaier and C. Nottnagel, *Elegantissimum ex physicis thema de hominibus apparenter mortuis* (Wittenberg, 1670).

21. G. M. Lancisi, *De subitaneis mortibus libri duo* (Rome, 1707), pp. 52–60. There is a good translation of this book under the title *On Sudden Deaths*, ed. P. D. White and A. V. Boursy (New York, 1971), containing the section on death signs on pp. 39–46. See also M. Hoffmann, *Die Lehre vom plötzlichen Tod in Lancisis Werk "De Subitaneis Morbis"* (Berlin, 1935).

22. F. Bacon, *Historia vitae et mortis* (London, 1623), pp. 431–33. Nothing about this sad ending of the life of the philosopher is mentioned in the various biographies of him; e.g., C. R. S. Harris, *Duns Scotus* (Oxford, 1927), merely states that he expired in November 1308 and was buried in the church of the friars minor in Cologne.

23. [Anon.], *The Most Lamentable and Deplorable Accident Which on Friday Last, June 22, Befell Laurence Cawthorne . . .* (London, 1661).

24. [Anon.], *A Full and True Relation of a Maid Living in Newgate-Street in London . . .* (London, 1680).

25. [Anon.], *News from Basing-Stoak* (London, 1674).

26. J. Jefferson, *The History of Holy Ghost Chapel, Basingstoke* (Basingstoke, 1819), pp. 84–88.

27. W. Tebb and E. P. Vollum, *Premature Burial and How It May Be Prevented* (London, 1905), p. 83. See also the article by I. A. Girvan in *Hampshire Magazine* 36 (1996): 23.

28. On early wills with precautions against premature burial, see Ariès, *Hour of Our Death*, pp. 399–401, and the article by J.-L. Bourgeon in *Revue d'Histoire Moderne et Contemporaine* 30 (1983): 139–53.

29. Bourgeon in *Revue d'Histoire Moderne et Contemporaine* 30 (1983): 140–41.

30. S. Goulart, *Thresor d'Histoires admirables et mémorables de nostre temps* (Geneva, 1620).

31. P. Zacchia, *Questiones medicolegales* (Amsterdam, 1651). The section on apparent death is in book 4, chap. 1, quest. 11, "De mortuorum resurrectione," pp. 241–50.

32. I. van Diemerbroeck, *Tractatus de peste* (Amsterdam, 1665), book 4, obs. 85.

33. On the Marseilles plague epidemic, see C. Carrière et al., *Marseille, ville morte* (Marseille, 1980).

34. On the pin implantation, see the articles by M. Signoli et al. in *Comptes Rendus de l'Académie des Sciences de Paris*, ser. 2a, 322 (1996): 333–39, and G. Leonetti et al. in *Journal of Forensic Sciences* 42 (1997): 744–48.

35. See W. G. Bell, *The Great Plague in London, 1665* (London, 1951), pp. 183–84, and P. Slack, *The Impact of the Plague in Tudor and Stuart England* (London, 1985), pp. 274–75.

36. The tale of the butcher is in the British Library, Dept. of Manuscripts, Add. MSS. 4182 f 39.

37. D. Defoe, *A Journal of the Plague Year* (London, 1970), pp. 107–8.

38. W. Austin, *Anatomy of the Pestilence* (London, 1665), p. 38.

CHAPTER 2: THE LADY WITH THE RING AND THE LECHEROUS MONK

1. *Notes and Queries*, 6th ser., 6 (1882): 209–11.

2. On the Richmodis von Aducht legend, see E. Weyden, *Cöln's Vorzeit* (Cologne, 1826), pp. 192–94, F. Bender, *Illustrierte Geschichte der Stadt Köln* (Cologne, 1924), pp., 129–30, G. P. Gath, *Kölner Sagen, Legenden und Geschichten* (Cologne, 1939), pp. 214–17, and C. Hinte, *Den alte Köln in Sagen* (Cologne, 1986), pp. 127–29. See also the articles by F. Hendriks in *Notes and Queries*, 6th ser., 4 (1881): 344–45, and W. E. A. Axon, ibid., pp. 518–19.

3. M. Misson, *A New Voyage to Italy* (London, 1695), 1: 40–41.

4. J. C. Risbeck, *Briefe eines reisenden Franzosen über Deutschland* (Zürich, 1784), 2: 496.

5. F. Bock, *Das heilige Köln* (Cologne, 1858), p. 8.

6. H. Merlo, *Die Famile Hackeney zu Köln* (Cologne , 1863), pp. 46–52.

7. For a critical analysis of the legend, see the article by J. Bolte in *Zeitschrift des Vereins für Volkskunde* 20 (1910): 359–62, and also the brief note by R. Dieckhoff in *KölnEdition*, vol. 10. I thank Dr. Jost Rebentisch, of the Kölnisches Stadtmuseum, for providing some original material.

8. The three articles by J. Bolte in *Zeitschrift des Vereins für Volkskunde* 20 (1910): 353–81, 21 (1911): 282–85, and 30–32 (1920–21): 127–30, provide an exhaustive overview of the German literature on the Lady with the Ring. I. Stoessel, *Scheintod und Todesangst* (Cologne, 1983), pp. 42–45, and C. Milanesi, *Mort apparente, mort imparfaite* (Paris, 1991), pp. 42–44, add some further valuable details.

9. On the countess of Mount Edgcumbe, see the article by Mrs. Bray in *Gentleman's Magazine*, n.s. 40 (1853): 444–50, and W. Tebb and E. P. Vollum, *Premature Burial and How It May Be Prevented* (London, 1905), p. 46.

10. R. Wilkins, *The Bedside Book of Death* (London, 1990), pp. 32–37.

11. The poem by R. S. Hawker in his *Records of the Western Shore* (1832), p. 13, is discussed in the article by W. E. A. Axon in *The Reliquary* 8 (1867–68): 146–50.

12. On other English Ladies with Rings, see ibid. and the articles by W. E. A. Axon in *Notes and Queries*, 6th ser., 4 (1881): 518–19, J. Pickford, ibid., vol. 5 (1881): 117–18, and H. Tripp, ibid., vol. 5 (1881): 159, and also Tebb and Vollum, *Premature Burial*, pp. 380–84, Wilkins, *Bedside Book of Death*, pp. 32–37, and R. Davies, *The Lazarus Syndrome* (London, 1999), pp. 150–51.

13. These two Scottish instances are detailed by K. V. Iserson, *Death to Dust* (Tucson, 1994), pp. 31–32.

14. P. Sieveking, "Misidentification of the Dead and Premature Burial, 1991–1997" (MS), p. 6.

15. H. Creighton, *Bluenose Ghosts* (Toronto, 1957), quoted in Davies, *Lazarus Syndrome*, pp. 152–53.

16. K. Lithner in *DAST-Magazine* 21, no. 4 (1988): 4–10.

17. See the articles by H. Schirmer in *Tidsskrift for den Norske Lœgeforeningen* 94 (1974): 2135–39, and by M. H:son Holmdahl in *Svenska Läkartidningen* 74 (1977): 3437–39, and in *Saga och Sed* 11 (1990): 35–43, and V. Starcke, *Giertrud Birgitte Bodenhoffs Mysterium* (Copenhagen, 1954).

18. On these curious legends, see J. Bondeson and A. Molenkamp, *The Prolific Countess* (Loosduinen, 1996), and J. Bondeson, *The Two-Headed Boy and Other Medical Marvels* (Ithaca, 2000), pp. 64–119.

19. On the legend of the Two Young Lovers, see H. Hachette, *La morte vivante* (Paris, 1933).

20. The Lecherous Monk is described by J.-J. Bruhier, *Dissertation sur l'incertitude des signes de la mort* (Paris, 1749), pp. 74–79.

21. On the untrue accusation of Andreas Vesalius, see C. D. O'Malley, *Andreas Vesalius of Brussels*, 1514–1564 (Berkeley, 1964), pp. 304–6, and articles by G. Matheson Cullen in *Edinburgh Medical Journal* 13 (1914): 324–39, 388–400, and by G. Sarton in *Isis* 45 (1954): 131–44, which also gives details of other alleged historical victims of Careless Anatomists. Milanesi, *Mort apparente*, pp. 100–106, adds some further sources on the Vesalius legend. It is remarkable that quite a few modern writers have taken the Vesalius legend seriously: see in particular the articles by E. Bendiner in *Hospital Practitioner* 21, no. 2 (1986): 199–207, and by I. I. Lasky in *Clinical Orthopaedics* 259 (1990): 304–11, and the books by K. V. Iserson, *Death to Dust* (Tucson, 1994), p. 28, and C. Quigley, *The Corpse: A History* (Jefferson, 1996), p. 185.

22. A. Burggraeve, *Étude sur André Vésale* (Gand, 1841), pp. 37–48, quoting an obscure work by the Spaniard Hernandez Moréjon and Don Antonio Lorente's history of the Inquisition. See also M. Roth, *Andreas Vesalius Bruxellensis* (Berlin, 1892), pp. 273–78.

23. On Cardinal Espinosa, see the article by G. E. Mackay in *Popular Science Monthly*

16 (1880): 389–97. According to some early nineteenth-century stories, the Careless Anatomist wrought havoc among the ranks of the Vatican. A certain Cardinal Spinosa was declared to be dead by his physicians, but when his chest was opened during the embalming process, his heart began to beat again, and he was just able to grasp the anatomist's knife before he fell dead with a groan. Exactly the same story was told about the Cardinal Somaglia, whose death "in a most lamentable manner" was fixed to August 1837. Both of these stories were of course *formes frustes* of the original legend about Cardinal Diego di Espinosa. See Tebb and Vollum, *Premature Burial*, pp. 273–74.

CHAPTER 3: WINSLOW THE ANATOMIST AND BRUHIER THE HORROR MONGER

1. On the life of Jacques-Bénigne Winslow, see his *Autobiographie*, ed. V. Maar (Copenhagen, 1912), E. Snorrasson, *L'anatomiste J.-B. Winslow* (Copenhagen, 1969), and the articles by E. W. Adams in *Medical Library and Historical Journal* 2 (1904): 28–34, and C. Gysel in *Histoire des Sciences Médicales* 19 (1985): 151–60.

2. See P. Schleisner, *Bidrag til Belysning af Asphyxien og Døden* (Copenhagen, 1868), p. 10.

3. This is my own translation from the Latin, which is more felicitous that the translation via the French in the 1746 English translation of Winslow's thesis; all later quotations from Winslow and Bruhier will, however, be from the 1746 London edition of the *Uncertainty of the Signs of Death*.

4. These cases are detailed in *Uncertainty of the Signs of Death*, pp. 6–9.

5. The tale of Philippe Peu is in *Uncertainty of the Signs of Death*, pp. 5–6, and is discussed by C. Milanesi, *Mort apparente, mort imparfaite* (Paris, 1991), pp. 108, 219. It should be noted, however, that E. Bouchut, *Traité des signes de la mort* (Paris, 1849), pp. 10–11, actually stated that Philippe Peu *almost* made the fatal incision, but stopped when he observed some slight sign of life, and that this version is supported by Peu's own cryptic note of the case in his *La pratique des accouchemens* (Paris, 1694), p. 334.

6. It is also hinted by Snorrasson, *L'anatomiste J.-B. Winslow*, p. 63, that Winslow's interest in the signs of death was one of long standing.

7. See Schleisner, *Bidrag til Belysning af Asphyxien og Døden*, p. 9.

8. For Winslow's ideas on the resuscitation of corpses, see *Uncertainty of the Signs of Death*, pp. 21–24.

9. Not much is known about the life of Jean-Jacques Bruhier, except for the article in the *Nouvelle biographie générale*, vol. 7 (Paris, 1855), pp. 586–87, which includes a bibliography, and the additional sources listed by Milanesi, *Mort apparente*, p. 208. It should be noted that Bruhier's role in the eighteenth-century debate on apparent death has been underestimated by some historians, who have misinterpreted his relation to Winslow. In particular, M. Alexander in *Hastings Center Reports* 10 (1980): 25–31, only referred to Bruhier as Winslow's student and translator. M. S. Pernick, in R. M. Zaner, ed., *Death: Beyond Whole-Brain Criteria* (Dordrecht, 1988), pp. 17–74, and S. M. Quinlan in *French History* 9 (1995): 27–47, also overestimate Winslow's contribution. In fact, with regard to its arguments, Winslow's thesis on the uncertainty of the signs of death was not far removed from the works of Zacchia and Lancisi. In contrast, P. Vecchi in *Revue de Synthèse*, ser. 3, 113–14 (1984): 143–60, and Milanesi, *Mort apparente*, pp. 13–52, have a correct appreciation of Bruhier's role.

10. See the preface to the 1749 edition of vol. 1 of Bruhier's *Dissertation sur l'incertitude des signes de la mort*, pp. 8–9, for this valuable addition to the publishing history of his book.

11. Bruhier's *Additions* begin on p. 34 of *The Uncertainty of the Signs of Death*; his treatise on funereal ceremonies is on pp. 133–202.

12. For the discussion of Kornmann's treatise and the cases from Dr. Crafft, see ibid., pp. 65–68 and 53–57.

13. For the tale of the holy water, see ibid., pp. 55–56; the scoffing remark about Lazarus is in the 1749 edition of Bruhier's *Dissertation sur l'incertitude des signes de la mort*, pp. 522–53.

14. The instances from Durham's book are discussed in *The Uncertainty of the Signs of Death*, pp. 57–63.

15. Bruhier's *Mémoire présenté au Roi* is reproduced in his *Dissertation sur l'incertitude des signes de la mort* (1749 ed.), pp. 547–88.

16. Bruhier himself reproduced some of the most flattering comments, ibid., pp. xxxix–l.

17. P.-F. Guyot Desfontaines, in *Observations sur les écrits modernes* 31 (Paris, 1742).

18. J.-J. Bruhier, *Dissertation sur l'incertitude des signes de la mort*, vol. 2 (Paris, 1746). His response to Desfontaines is on pp. 17–94; his lengthy treatise on submarine Swedes, on pp. 95–221.

19. M. de la Sorinière's *Éloge* is reproduced in Bruhier, *Dissertation sur l'incertitude des signes de la mort*, 1749 ed., pp. xxxvii–xxxviii.

20. Ibid., pp. 608–9.

21. See *Uncertainty of the Signs of Death*, pp. 42–45.

22. See ibid., pp. 6–7 (Orléans), 41–42 (Poitiers), 42–45 (Cologne: von Aducht case), and 69–70 (Toulouse); also *Dissertation sur l'incertitude des signes de la mort*, 1749 ed., pp. 28–29 (Orléans), 53–55 (Poitiers), 61–62 (Toulouse), 70–72 (Bordeaux), 98–100 (Dublin), 134–38 (von Aducht case), 170–73 (Leipzig).

23. Bruhier, *Dissertation sur l'incertitude des signes de la mort* (1749 ed.), pp. 69–74 (Pitaval's story), 94–97 (Périgord).

24. See ibid., p. 49.

25. Ibid., pp. 107–10.

26. On Count Richard, see ibid., pp. 77–78.

27. Ibid., pp. 84–86 (Dole case) and 537–46 (Rigadeux).

28. *Gentleman's Magazine* 15 (1746): 260.

29. [Anon.], *The Uncertainty of the Signs of Death* (London, 1746). Bruhier was aware of the English edition and quoted the Basingstoke case from it in his *Dissertation sur l'incertitude des signes de la mort* (1749 ed.), pp. 104–6.

30. [Anon.], *The Uncertainty of the Signs of Death* (London, 1751), pp. iii–v.

31. J. B. Winslow, *Afhandling om Dödstecknens Ovisshet* (Stockholm, 1751); for the additions about the two Swedish instances, see pp. 172–78, 189–90.

32. See the article by P. Hedenius in *Upsala Läkareförenings Förhandlingar* 10 (1874–75): 31–35.

33. J.-J. Bruhier, *Abhandlung von der Ungewißheit der Kennzeichen des Todes*, trans. J. G. Jancke (Leipzig, 1754).

CHAPTER 4: THE EIGHTEENTH-CENTURY DEBATE

1. On Antoine Louis's life and his pioneering work in forensic medicine, see M. Silie, *Un des promoteurs de la médicine légale française, Antoine Louis* (Paris, 1924), and J. Bourakhovitch, *Contribution à la bio-bibliographie d'Antoine Louis* (Rennes, 1958), and the articles by C. B. Courville in *Bulletin of the Los Angeles Neurological Society* 10 (1945): 46–69, P. Astruc in *Progrès Médical* 84 (1956): 227, and K. Egeblad in *Dansk Medicinhistorisk Årbog* 17 (1988): 29–44.

2. P. J. B. Previnaire, *Abhandlung über die verschiedenen Arten des Scheintodes* (Leipzig, 1790), p. 44, and C. C. Creve, *Vom Metallreize* (Leipzig, 1790), pp. 24–25.

3. Jancke, in his preface to the *Abhandlung von der Ungewißheit der Kennzeichen des Todes* (Leipzig, 1754).

4. Both Bruhier's attack and Louis's rebuttal were in the *Mercure de France* of September 1752, pp. 210–12.

5. For some French examples, see E.-M.-L. Dosias, *Des signes de la mort* (Paris, 1858), p. 7, and A.-L.-P. Bonnard, *Des signes de la mort apparente et de la mort réelle* (Montpellier, 1844), pp. 5–6; an English one is given in an anonymous article in the *London Medical Record* 2 (1874): 205–7. M. Augener, *Scheintod als medizinisches Problem im 18. Jahrhundert* (Kiel, 1965), is another example of this attitude.

6. J.-L. Bourgeon in *Revue d'Histoire Moderne et Contemporaine* 30 (1983): 139–53.

7. On the process of dechristianization, see M. Vovelle, *Religion et révolution* (Paris, 1976). Its effect on the debate on apparent death has been discussed by P. Ariès, *The Hour of Our Death* (New York, 1981), pp. 399–401, R. Favre, *La mort dans la littérature et la pensée françaises au siècle des lumières* (Lyon, 1978), pp. 265–71, and the valuable articles by P. Vecchi in *Revue de Synthèse* ser. 3, 113–14 (1984): 143–60, and S. M. Quinlan in *French History* 9 (1995): 27–47.

8. Ariès, *Hour of Our Death*, pp. 205–6, 234–37.

9. B. Puckle, *Funeral Customs* (London, 1926), pp. 42–43.

10. F. P. Wilson, *The Plague in Shakespeare's London* (Oxford, 1927), p. 43.

11. On the "Toad in the Hole" mystery and its various implications, see J. Bondeson, *The Feejee Mermaid and Other Essays* (Ithaca, 1999), pp. 280–308. J.-J. Bruhier discussed the extreme hardiness of amphibians in his *Dissertation sur l'incertitude des signes de la mort* (Paris, 1746), 2:142–43. Le Cat's article was in Alléon Dulac's *Mélanges d'Histoire Naturelle* 3 (1764): 95–105.

12. Bruhier, *Dissertation sur l'incertitude des signes de la mort*, 2:116–36.

13. On "fasting girls," see G. M. Gould and W. L. Pyle, *Anomalies and Curiosities of Medicine* (New York, 1956), pp. 414–21, and also the papers by H. E. Rollins in *Journal of American Folklore* 34 (1921): 357–76, W. Vandereycken and R. Van Deth in *History Today*, no. 8 (1993): 37–39, and Ricky Jay in *Jay's Journal of Anomalies* 4, no. 1 (1998).

14. See the papers by A. Hj. Uggla in *Lychnos* (1940), pp. 302–24, and M. H:son Holmdahl in *Saga och Sed* 11 (1990): 35–43.

15. In the *Philosophical Transactions of the Royal Society of London* 2 (1667): 539.

16. R.-A. de Réaumur, *Avis pour donner du secours à ceux que l'on croit noyés* (Paris, 1740). It was pp. 344–56 in the 1742 edition of Bruhier's book.

17. [Anon.], *The Uncertainty of the Signs of Death* (London, 1751), p. v.

18. See Favre, *La mort dans la littérature*, pp. 266–69, and C. Milanesi, *Mort apparente, mort imparfaite* (Paris, 1991), pp. 123–37. Buffon wrote on apparent death in the sec-

ond volume of his *Histoire naturelle de l'homme* (Paris, 1749), republished with additions in 1777.

19. Originally titled *Terrible supplice et cruel désespoir des personnes enterrées vivantes et qui sont présumées mortes* (Paris, 1752), this pamphlet contains mainly a French translation of Winslow's original thesis.

20. See F. Lebrun, *Les hommes et la mort en Anjou* (Paris, 1971), pp. 460–61.

21. On the fear of mephitic poisoning in France at this time, see Milanesi, *Mort apparente*, pp. 154–57.

22. On the abbé Prévost's supposed encounter with the Careless Anatomist, see H. Harrisse, *L'abbé Prévost* (Paris, 1896), pp. 427–53; also A. Billy, *Un singulier bénédictin: L'abbé Prévost* (Paris, 1969), pp. 303–7. A similarly lurid tale was spread about the actress Elizabeth Rachel Felix, better known as Mlle Rachel. One of the untrue posthumous rumors about her was that after the process of embalming her body had begun, she had actually awakened from her deep death trance, but died ten hours afterward from the injuries inflicted on her, in the horrid agonies that narrow-minded people thought befit her immoral life. See F. Hartmann, *Premature Burial* (London, 1896), p. 80. For true accounts of her life and death, which mention nothing about these dramatic excesses, see J. Richardson, *Rachel* (London, 1956), and B. Falk, *Rachel the Immortal* (Bath, 1974). Hartmann also cites an American newspaper story from 1864, in which the "corpse" jumped up from the dissection table and grasped the Careless Anatomist by the throat. The conclusion, much relished by anti-premature-burial propagandists of the time, was that the anatomist died of apoplexy on the spot and that the "corpse" recovered fully.

23. See the article by E. Lesky in *Wiener klinische Wochenschrift* 84 (1972): 244–45.

24. J. P. Brinckmann, *Beweis der Möglichkeit, daß einige Leute lebendig können begraben werden* (Düsseldorf, 1772).

25. On this debate, see M. Herz, *Über die frühe Beerdigungen der Juden* (Berlin, 1788), and the later papers by F. Wiesemann in *Leo Baeck Institute Yearbook* 37 (1992): 17–31, and J. M. Efron in *Bulletin of the History of Medicine* 69 (1995): 349–66.

26. See the paper by P. Pasture in *Tijdschrift voor Geschiedenis* 100 (1987): 198–217.

27. On the foundation of humane societies in the 1760s and 1770s, see the article by E. H. Thomson in *Bulletin of the History of Medicine* 37 (1963): 43–51.

28. See the article by G. Puppe in *Deutsche medizinische Wochenschrift* 46 (1920): 383–85.

29. Those by J. Janin de Combe-Blanche, *Réflexion sur le triste sorte des personnes, qui sous une apparence de mort, ont été enterrées vivantes* (The Hague, 1772), J.-J. de Gardanne, *Avis du peuple sur les morts apparentes et subites* (Paris, 1774), and Dr. Pineau, *Mémoire sur le danger des inhumations précipitées* (Paris, 1776).

30. W. Hawes, *An Address to the Public on Premature Death and Premature Interment* (London, 1780), p. 40.

31. Quoted in C. Gittings, *Death, Burial and the Individual in Early Modern England* (London, 1984), p. 205.

32. On Miss Beswick, see Edith Sitwell, *The English Eccentrics* (London, 1947), pp. 22–23, and the articles by T. K. Marshall in *Medicolegal Journal* 35 (1966): 14–24, and B. Haworth in *Udolpho* 18 (1994): 35.

CHAPTER 5: HOSPITALS FOR THE DEAD

1. F. Thiérry, *La vie de l'homme respectée et defendue dans ses derniers moments* (Paris, 1787). It was translated into German as *Unterricht von den Fürsorge, die man den Todten schuldig ist* (Lübeck, 1788).

2. On Johann Peter Frank and his subsequent career, see H. Haubold, *Johann Peter Frank* (Munich, 1939), and the articles by L. Baumgartner and E. M. Ramsey in *Annals of Medical History*, n.s. 5 (1933): 525–32, and n.s. 6 (1934): 69–90. The section on apparent death and waiting mortuaries is in his *System einer vollständigen medicinischen Polizey*, vol. 4 (Mannheim, 1788), pp. 672–749.

3. See E. Lesky, ed., *Johann Peter Frank: Seine Selbstbiographie* (Stuttgart, 1969), p. 63.

4. On Hufeland and his distinguished career, see K. Pfeifer, *Christoph Wilhelm Hufeland: Mensch und Werk* (Halle, 1968), W. Genschorek, *Christoph Wilhelm Hufeland* (Leipzig, 1976), and H. Busse, *Christoph Wilhelm Hufeland* (St. Michael, 1982); there are few English-language sources, although the article by I. A. Abt in *Annals of Medical History*, n.s. 3 (1931): 27–38, is recommended. Already at the age of twenty-one, Hufeland had presented a doctoral thesis on the use of electricity to save apparently dead people.

5. In his autobiography, Hufeland stated that he had been inspired by Frank's appeal. See W. von Brunn, ed., *Hufeland, Leibarzt und Volkserzieher* (Stuttgart, 1937), p. 70.

6. The collection is described by Pfeifer, *Christoph Wilhelm Hufeland*, p. 71.

7. C. W. Hufeland, *Ueber die Ungewißheit des Todes und das einzige untrügliche Mittel sich von seiner Wirklichkeit zu überzeugen: Nebst der Nachricht von der Errichtung eines Leichenhaus in Weimar* (Weimar, 1791).

8. The building program for German waiting mortuaries has been described by H.-K. Boehlke, *Wie die Alten den Tod gebildet* (Mainz, 1979), pp. 135–146, and more exhaustively by M. U. Stein, *Das Leichenhaus* (Marburg/Lahn, 1992).

9. On the first Berlin waiting mortuary, see Stein, *Das Leichenhaus*, p. 26, and T. Koch, *Lebendig begraben* (Augsburg, 1996), p. 101.

10. On the first Munich waiting mortuary and contemporary opinion about its various defects, see C. Schwalbe, *Das Leichenhaus in Weimar* (Leipzig, 1834), pp. 13–15.

11. [Anon.], *Wiederauflebungs-Geschichten scheintodter Menschen* (Berlin, 1798). It was translated into Swedish under the title *Historier om Skendöde som åter upwaknat* (Stockholm, 1800). The stories about the monk, the Portuguese old man, and the inscription in the vault are on pp. 68–71, 60–64, and 57–59, respectively, in the Swedish edition.

12. H. F. Köppen, *Achtung des Scheintodtes*, 2 vols. (Halle, 1800). This work utilizes the same set of stories as the *Wiederauflebungs-Geschichten scheintodter Menschen*, but adds six new ones.

13. H. v. C., *Wirkliche und wahre mit Urkunden erläuterte Geschichten und Begebenheiten von lebendig begrabene Personen, welche wiederum aus Sarg und Grab erstanden sind* (Frankfurt and Leipzig, 1798). The stories in this obscure collection are discussed by S. Schäfer, *Scheintod: Auf den Spuren alter Ängste* (Berlin, 1994).

14. C. W. Hufeland, *Der Scheintod* (Berlin, 1808). A facsimile reprint of this book, edited and interestingly prefaced by G. Köpf, was published in 1986 by P. Lang Publishers.

15. See D. C. Bastholm, *En Opfordring til Kiøbenhavns Indvaanere . . .* (Copenhagen, 1793).

16. A similarly macabre example of this perverted zeal for philanthropy was the debate over whether the head of a decapitated person could live, and feel, for a prolonged

period of time. The philanthropists urged that the still-living, still-feeling severed head must suffer unimaginable torments; surely, this made decapitation the most dreadful and barbaric mode of execution. Hufeland and other physicians undertook gruesome experiments with the severed heads of beheaded criminals to support their point. See Koch, *Lebendig begraben*, pp. 81–88, and the article by L. Jordanova in *History Workshop Journal* 28 (1989): 39–52.

17. J. D. Metzger, *Ueber die Kennzeichen des Todes* (Königsberg, 1792).

18. Von Müller, *Wie sich lebendig Begrabene gar leicht wieder aus Sarg und Grab helfen und ganz bequem herausgehen können* (Prague, 1790); it is reviewed in Hufeland's *Der Scheintod*, pp. 230–32.

19. Hufeland, *Der Scheintod*, pp. 33–34, 177–79, for horror stories from the *Wiederauflebungs-Geschichten scheintodter Menschen*; pp. 5–7, 18–20, 108–9, 134–35, 173–75, for various ludicrous stories from Waginger's *Neue Gespenster* .

20. J. Atzel, *Über Leichenhäuser vorzüglich als Gegenstand der schönen Baukunst betrachtet* (Stuttgart, 1796). This book is interestingly discussed by Boehlke, *Wie die Alten den Tod gebildet*, pp. 135–46.

21. The architectural styles of the German waiting mortuaries are well described by Stein, *Das Leichenhaus*.

22. On the first Frankfurt waiting mortuary, see M. B. Lessing, *Skendöden eller det Osäkra i wår Kännedom om Lifwets Utslocknande* (Linköping, 1837), pp. 137–41.

23. Schwalbe, *Das Leichenhaus in Weimar*.

24. L. A. Kraus, *Das Sterben im Grabe* (Helmstedt, 1837). On the lousy disease, see my *Cabinet of Medical Curiosities* (Ithaca, 1997), pp. 51–71.

25. J. C. A. Clarus in *Beiträge zur praktischen Heilkunde* 1 (1834): 532–35.

26. For these five indications of lack of faith in the German *Leichenhäuser*, see, respectively, C. L. Klose in *Zeitschrift für die Staatsarzneikunde* 19 (1830): 143–75; E. Burkel, *Über die Verhütung des Scheintodes* (Berlin, 1984), p. 38; A. van Hasselt, *Die Lehre vom Tode und Scheintode* (Brunswick, 1862), pp. 75–83; J.-A. Josat, *De la mort et de ses caractères* (Paris, 1854); and Dr. Graff in *Zeitschrift für die Staatsarzneikunde* 34 (1837): 273.

27. E. G. von Steudel, *Altbau und Neubau des Medizinal-Wesens in Württemberg* (Esslingen, 1849). This work was later commented on by D. Gross in *Würzburger medizinhistorische Mitteilungen* 16 (1997): 15–33.

28. These details are given by B. Gaubert, *Les chambres mortuaires* (Paris, 1895), pp. 132–38.

29. The Frankfurt mortuary rebuilt in 1848 was well described by Josat, *De la mort et de ses caractères*, pp. 170–201, and later by Gaubert, *Les chambres mortuaires*, pp. 138–40.

30. G. Le Bon, *Om Skindød og om Forhastede Begravelser* (Copenhagen, 1867), pp. 95–98.

31. Gaubert, *Les chambres mortuaires*, pp. 129–31.

32. The Munich waiting mortuary was well described by W. Tebb and E. P. Vollum, *Premature Burial and How It May Be Prevented* (London, 1905), pp. 341–49.

33. W. Collins, *Jezebel's Daughter* (London, 1880).

34. This article is quoted by Dr. Moore Russell Fletcher in *One Thousand People Buried Alive by Their Best Friends* (Boston, 1883), p. 40.

35. S. L. Clemens, *The Complete Short Stories of Mark Twain* (New York, 1957), pp. 225–34. The story "A Dying Man's Confession" was originally in his *Life on the Mississippi*.

36. E. Vogl, *Der Scheintod, eine medizingeschichtliche Studie* (Munich, 1986), p. 19.

37. See the article by G. Rossow in *Beiträge zur gerichtlichen Medizin* 17 (1943): 125. An article in *Der Spiegel* (1967, no. 48), p. 177, confirms that certain German mortuaries had signaling contraptions as late as the 1940s and 1950s. According to Schäfer, *Scheintod*, p. 80, the Strasbourg crematorium still had alarm buttons in the refrigerated boxes used for storing corpses in the 1960s.

38. S. Necker, *Les inhumations précipitées* (Paris, 1790). Mme Necker lived in dread of being buried alive for the remainder of her days. In her will, she implored her husband that she never be buried or put in a coffin. Her embalmed body was put in a large bathtub full of alcohol, which was covered by a glass case and put in a secure vault. Ten years later, Jacques Necker's own corpse was treated in exactly the same manner, and the door to their mausoleum was walled up. Since that time, it has been opened only once: to admit the coffin of the Neckers' daughter Germaine, better known as the literary hostess Mme de Staël. Some modern historians, like P. Vecchi in *Revue de Synthèse*, ser. 3, 113–14 (1984): 143–60, and M. S. Pernick, *Death: Beyond Whole-Brain Criteria*, ed. R. M. Zaner (Dordrecht, 1988), pp. 17–74, have exaggerated the importance of Mme Necker's pamphlet, claiming that it influenced Hufeland and subsequently the German waiting mortuaries. Milanesi, *Mort apparente*, pp. 187–91, is aware of Thiérry's role, but still overstates the importance of the Necker pamphlet. This pamphlet is mentioned nowhere in Hufeland's 1791 pamphlet and is quite unlikely to have influenced him, particularly when the correct publication history of his early works on waiting mortuaries, as presented by Stein, *Das Leichenhaus*, p. 22, is considered. Moreover, Schwalbe, who knew and admired Hufeland, admitted in his *Leichenhaus in Weimar*, p. 10, the latter's debt to Frank's work.

39. L. von Berchtold, *Kurzgefaßte Methode, alle Arten von Scheintodten wieder zu beleben* (Vienna, 1791), published in French under the title *Projet pour prévenir les dangers très frequens des inhumations précipitées* (Paris, 1791). There was a Spanish translation in 1792.

40. These appeals to build waiting mortuaries in Paris have been chronicled by Gaubert, *Les chambres mortuaires*, pp. 56–80.

41. Bastholm, *En Opfordring til Kiøbenhavns Indvaanere.*

42. The Copenhagen waiting mortuary is described by P. A. Schleisner, *Bidrag til Belysning af Asphyxien og Döden* (Copenhagen, 1868), pp. 65–66, and by V. Starcke, *Giertrud Birgitte Bodenhoffs Mysterium* (Copenhagen, 1954).

43. For examples of Scandinavian pro-mortuary propaganda, see the pamphlets of N. L. Nissen, *Om Skindød, levendes Begravelse og Lighuse til Forebyggelse heraf* (Copenhagen, 1827), and [Anon.], *Uppmaning till Menniskovennen . . .* (Uddevalla, 1840).

44. See A. V. Zarda, *Patriotischer Wunsch für die Wiederbelebung vertodtscheinenden Menschen* (Prague, 1797).

45. The Vienna waiting mortuary was described in the *Illustrirtes Wiener Extrablatt*, October 5, 1874, and later by H. Schmölzer, *A Schöne Leich: Der Wiener und sein Tod* (Vienna, 1980), pp. 95–100, and E. Burkel, *Über die Verhütung des Scheintodes* (Berlin, 1984), pp. 33–36. S. Jellinek, *Dying, Apparent Death, and Resuscitation* (London, 1947), p. 37, describes how he took the alarm system from the old mortuary of a Vienna hospital to his private collection in 1899.

46. See Josat, *De la mort et de ses caractères*, pp. 171–72.

47. [Anon.], *Iets over het Levend-begraven en de Lijkenhuizen* (Haarlem, 1837), p. 31.

48. These details are given by E. Bouchut, *Traité des signes de la mort et des moyens*

de prévenir les inhumations prématurées (Paris, 1883), pp. 410–14; see also Gaubert, *Les chambres mortuaires*, p. 155.

49. On the old *Schijndodenhuis* of The Hague, see the articles by E. M. Terwen-Dionysius in *Jaarboek die Haghe* (1986): 67–99, R. Spruit in *Spieghel Historiael* 21 (1986): 444–49, and C. van Raak in *Maatstaf* 42, no. 2 (1994): 32–38.

50. Stein, *Das Leichenhaus*, pp. 141, 143.

51. Froriep's *Notizen aus dem Gebiete der Natur- und Heilkunde* 24 (1829): 255.

52. R. Brandon in *Medical Times* 16 (1847): 574, and [anon.] in *British and Foreign Medico-Chirurgical Review* 15 (1855): 74.

53. I here use the word "damned" like that ingenious philosopher and student of macabre phenomena—Charles Fort—to signify an unexplained phenomenon deliberately overlooked, or a piece of history thought to be best left forgotten.

54. On Varrentrapp's appeal, see van Hasselt, *Die Lehre vom Tode und Scheintode*, p. 65.

55. L. Lenormand, *Des inhumations précipitées* (Macon, 1843), p. 149.

56. Le Bon, *Om Skindød og om Forhastede Begravelser*, pp. 95–98.

57. Dr. Hofmann in *Allgemeine medicinische Central-Zeitung* 16 (1847): 612–14.

58. Josat, *De la mort et de ses caractères*, pp. 204–7.

59. Le Bon, *Om Skindød og om Forhastede Begravelser*, pp. 95–98.

60. Gaubert, *Les chambres mortuaires*, pp. 178–81.

61. Tebb and Vollum, *Premature Burial*, pp. 348–49.

62. See van Hasselt, *Die Lehre vom Tode und Scheintode*, pp. 73–74.

63. Brouardel, *Death and Sudden Death*, p. 23; see also Vogl, *Der Scheintod*, p. 19.

CHAPTER 6: SECURITY COFFINS

1. The duke's security coffin is described by C. W. Hufeland, *Der Scheintod* (Berlin, 1808), p. 93.

2. For reviews of the German obsession with security coffins, see H. Schmid, *Historische Analyse des Scheintodes* (Munich, 1983), pp. 55–66 and appendixes; also E. Burkel, *Über die Verhütung des Scheintodes* (Berlin, 1984), pp. 44–70.

3. P. G. Pessler, *Leicht anwendbarer Beystand der Mechanik, um Scheintodte beym Erwachen im Grabe zu erretten* (Brunswick, 1798).

4. Collenbusch's *Rath-geber für aller Stände*, vol. 1 (1799), and Poppe's *Allgemeines Rettungsbuch* (Hanover, 1805), pp. 495–514, 533–46, both quoted in Hufeland, *Der Scheintod*, pp. 8–9, 198–99.

5. J. G. Hypelli, *Ein Wecker auch Rettungsmittel für Scheintodte* (Burghausen, 1804), discussed by E. Burkel, *Über die Verhütung des Scheintodes* (Berlin, 1984), pp. 44–46.

6. This idea and others were discussed in Poppe's *Noth- und Hülfslexicon*, 1:338; see also the anonymous Dutch pamphlet *Iets over het Levend-begraven en de Lijkenhuizen* (Haarlem, 1837), pp. 28–29.

7. On the activities of Dr. Gutsmuth and Dr. von Hesse, see Schmid, *Historische Analyse des Scheintodes*, pp. 57–58.

8. J. G. Taberger, *Der Scheintod* (Hanover, 1829). This important book is discussed by Schmid, *Historische Analyse des Scheintodes*, pp. 58–62.

9. The account of the portable death chambers (*tragbare Todtenkammern*) is from M. U. Stein, *Das Leichenhaus* (Marburg/Lahn, 1992), pp. 19–20, which gives as sources an

article in an obscure publication, the Dresden *Abend-Zeitung* of 1829, and an account in Staatsarchiv Weimar B 7655.

10. These security devices are discussed by Schmid, *Historische Analyse des Scheintodes*, pp. 63–65, and by E. Burkel, *Über die Verhütung des Scheintodes* (Berlin, 1984), pp. 46–70; both of these thesis books reproduce original patent applications and drawings for the coffins involved.

11. Bateson's Belfry is described by M. Crichton, *The Great Train Robbery* (London, 1975), pp. 187–93, and L. G. Stevenson in *Bulletin of the History of Medicine* 49 (1975): 482–511.

12. See *Burial Reformer* 3 (1911): 23.

13. For an overview of U.S. security coffins, see the article by H. Dittrick in *Bulletin of the History of Medicine* 3 (1948): 161–71.

14. This amusing account is from the London *Times* of Sept. 16, 1868, p. 7c, and Sept. 29, 1868 , p. 7a.

15. The legend of Mary Baker Eddy's telephone is reviewed in an Internet homepage on premature burial (snopes.simplenet.com/horrors/gruesome/buried.htm). A similar story was current about another famous evangelist, Aimée McPherson. For the tale of Mrs. Pennord, see R. Davies, *The Lazarus Syndrome* (London, 1999), p. 241. According to an article in the *American Mercury* of 1926, the stockbroker Martin A. Sheets had electric lighting and a telephone installed in his mausoleum.

16. See the articles by H. Valbel in *Journal d'Hygiène* (Paris), 22 (1897): 106–8, and Dr. de Pietra Santa, ibid., pp. 157–63.

17. M. de Karnice-Karnicki, *Vie ou mort* (Paris, 1900), pp. 28–31. On the count's Italian activities, see also *Ingegneria sanitaria* 8 (1897): 121–22.

18. M. de Karnice-Karnicki, *Vie ou mort*. His associate Horace Valbel wrote another pamphlet, entitled *Der Scheintod besiegt durch "Le Karnice": Ein Werk der Menschlichkeit*, published in 1900 and later translated into Italian.

19. The article by E. Camis was in the *Medico-Legal Journal* (New York), 17 (1899–1900): 296–99.

20. N. R. Aronstam and L. J. Rosenberg in *Medical and Surgical Monitor* 4 (1901): 55–59.

CHAPTER 7: THE SIGNS OF DEATH

1. A. Louis, *Lettres sur la certitude des signes de la mort* (Paris, 1752), pp. 153–56.

2. For this story, see R. Wilkins, *The Fireside Book of Death* (London, 1990), p. 39.

3. For a review of the crude eighteenth-century methods of resuscitation, see Winslow and Bruhier's *The Uncertainty of the Signs of Death* (London, 1746), pp. 21–24.

4. D. G. Björn, *Medel at förekomma lefwande människors begrafning* (Linköping, 1795).

5. Louis, *Lettres sur la certitude des signes de la mort*.

6. P. J. B. Previnaire, *Abhandlung über die verschiedenen Arten des Scheintodes* (Leipzig, 1790). This is a German translation of a French edition that appeared a few years earlier.

7. R. A. Gorter, *De Tabaksrook-klisteer voornamelijk als Reanimator* (Amsterdam, 1953). This author has also written a series of articles about enemas of tobacco smoke, published in various Dutch periodicals, reprints of which are deposited in the British Library.

8. See Wilkins, *Fireside Book of Death*, pp. 43–44.

9. See the articles by D. C. Schechter in *Surgery* 69 (1971): 360–72, and N. Roth in *Medical Instrumentation* 14 (1980): 322. In 1788, Dr. Charles Kite suggested, in his *Essay on the Recovery of the Apparently Dead* (London, 1788), p. 125, that the electric stimulation of a muscle would be useful as a sign of death, for if the person was really dead, the muscle would not twitch.

10. C. C. Creve, *Vom Metallreize* (Leipzig, 1796).

11. A. V. Zarda, *Patriotischer Wunsch für die Wiederbelebung vertodtscheinenden Menschen* (Prague, 1797), p. 54.

12. C. W. Hufeland, *Der Scheintod* (Berlin, 1808), pp. 35–36.

13. C. A. Struwe, *Der Lebensprüfer* (Hanover, 1805). In 1804, Dr. J. A. Heidmann had presented a machine rather like that of Struwe, in a book entitled *Zuverlässiges Prüfungsmittel zur Bestimmung des wahren Scheintode* (Vienna, 1804). On other electrical devices for testing life, see also the typescript by M. Augener, "Scheintod als medizinisches Problem" (Kiel, 1965), pp. 74–82.

14. F.-M.-X. Bichat, *Recherches physiologiques sur la vie et la mort* (Paris, 1800), translated into English as *Physiological Researches upon Life and Death* (Philadelphia, 1809). It is curious that Bichat himself did not comment on the relevance of his work to the debate about the uncertainty of the signs of death, except to doubt the diagnostic value of cadaveric rigidity. Nor does E. Haigh, *Xavier Bichat and the Medical Theory of the Eighteenth Century*, Medical History Suppl. 14 (London, 1984).

15. See M. Orfila, *Secours à donner aux personnes empoisonnées ou asphyxiées* (Paris, 1818), and F. E. Fodéré, *Traité de médecine légale*, vol. 2 (Paris, 1813), pp. 343–73.

16. These theses are listed and reviewed by F. Gannal, *Mort réelle et mort apparente* (Paris,1868), pp. 26–30.

17. Pietro Manni had himself published a treatise on apparent death: *Manuale pratico per la cura degli apparentementi morte* (Naples, 1835). On the foundation and the inconclusive early contests for the Prix Manni, see the article by M. Rayer in *Comptes Rendus de l'Academie des Sciences* 26 (1848): 550–73.

18. R. Laënnec, *De l'auscultation médiate ou traité du diagnostic des maladies des poumons et du coeur* (Paris, 1819).

19. C. F. Nasse, *Die Unterscheidung des Scheintodes vom wirklichen Tode* (Bonn, 1841).

20. J. N. Hickmann, *Die Elektricität als Prüfungs- und Belebungsmittel im Scheintode* (Vienna, 1841).

21. The *pince-mamelon*, or nipple pincher, is described and depicted in J.-A. Josat, *De la mort et de ses caractères* (Paris, 1854).

22. The various other suggestions of signs of death or tests of life have been reviewed by T. K. Marshall in *Medicolegal Journal* 35 (1966): 14–24, and M. S. Pernick, *Death: Beyond Whole-Brain Criteria*, ed. R. M. Zaner (Dordrecht, 1988), pp. 17–74.

23. M. Rayer's report was published in the *Comptes Rendus de l'Academie des Sciences* 26 (1848): 550–73. For an example of its foreign reception, see the article by C. G. Santesson in *Hygiea* 10 (1848): 667–70.

24. E. Bouchut, *Traité des signes de la mort* (Paris, 1849); new editions appeared in 1854 and 1883. Eugène Bouchut was born in 1818 and died in 1891; he was one of many distinguished French physicians of this period.

25. See, e.g., M. Lassere in *Annales de Démographie Historique* (1992): 339–42.

26. See, e.g., R. Olry in *Vesalius* 2 (1996): 111–17.

27. T. Plugge in *Memorabilien* 5 (1860): 71.

28. W. B. Kesteven in *British and Foreign Medico-Chirurgical Review* 15 (1855): 71–77.

29. On the French opposition to Bouchut, see Gannal, *Mort réelle et mort apparente*, pp. 38–47, S. Icard, *La mort réelle et la mort apparente* (Paris, 1897), pp. 73–95, and J. B. Geniesse, *La mort réelle et la mort apparente* (Paris, 1905), pp. 392–94.

30. Icard, *La mort réelle et la mort apparente*, pp. 87–88.

31. Ibid., p. 45.

32. For the needle test of Cloquet and Laborde, see the article by Pernick, in *Death: Beyond Whole-Brain Criteria*, p. 39.

33. For the test of Marteno, see Wilkins, *Fireside Book of Death*, pp. 40–41.

34. U. Magnus, *Di un segno certo della morte* (Milan, 1879); see also A. Monteverdi, *Vita e Morte* (Cremona, 1892), pp. 146–47.

35. A. Monteverdi, *Note sur un moyen simple, facile, prompt et certain de distinguer la mort vrai de la mort apparente de l'homme* (Cremona, 1874).

36. L. Collongues, *Traité de dynamoscopie; ou, Appréciation de la nature et de la gravité des maladies par l'auscultation des doigts* (Paris, 1862), esp. pp. 348–70.

37. J.-V. Laborde, *Le traitement physiologique de la mort* (Paris, 1894) and *Le signe automatique de la mort réelle* (Paris, 1900). The method was reviewed in Geniesse, *La mort réelle et la mort apparente*, pp. 110–11.

38. This ill-deserved accolade, from the *Etudes Fransiscaines*, was quoted in Geniesse, *La mort réelle et la mort apparente*, p. 151.

39. The contest for the Prix d'Ourches was reviewed in the *London Medical Record* 2 (1874): 205–7, 221–23.

40. A. Devergie in *Annales d'Hygiène Publique* 34 (1870): 310–27.

41. See the *London Medical Record* 2 (1874): 205–7, 221–23, and the article by H. Carrington in *Annals of Psychical Science* 9 (1910): 255–67.

42. On the Prix Dusgate, see W. Tebb and E. P. Vollum, *Premature Burial and How It May Be Prevented* (London, 1905), pp. 312–13, 319.

43. Bouchut, *Traité des signes de la mort* (Paris, 1883).

44. Icard, *La mort réelle et la mort apparente*, pp. 92–93, 261–63.

45. Ibid. Dr. Icard was active for many years as one of the last serious French medical campaigners against premature burials: from 1897 until 1919, he wrote many books and articles about corpses, putrefaction, funerals, and autopsies. Born in 1860, he was still alive in 1932, when he published a book on heraldry.

46. See the article by L. A. Parry in *Medical Magazine* 23 (1914): 11–23.

CHAPTER 8: SKEPTICAL PHYSIOLOGISTS AND RAVING SPIRITUALISTS

1. J. Taylor, *The Danger of Premature Interment* (London, 1816).

2. W. Whiter, *A Treatise on the Disorder of Death* (London, 1817).

3. J. Snart, *Thesaurus of Horror; or, The Charnel-House Explored* (London, 1817). It was later reprinted under the more balanced title *A Historical Inquiry concerning Apparent Death and Premature Interment* (London, 1824); it is significant that this edition is much scarcer than the *Thesaurus of Horror*.

4. Snart, *Thesaurus of Horror*, pp. 95–96.

5. These examples are quoted by W. Tebb and E. P. Vollum, *Premature Burial and How It May Be Prevented* (London, 1905), pp. 186–89, 379; some additions are given by "Grime" in *Notes and Queries*, 3d ser., 2 (1862): 110, and by M. Pallot in *Udolpho* 26 (1996): 3–5.

6. See N. Longmate, *King Cholera* (London, 1966), p. 54, R. J. Morris, *Cholera 1832* (London, 1976), pp. 105–7, and M. Durey, *The Return of the Plague* (Dublin, 1979), pp. 168–69.

7. This account is from the London *Times*, May 4, 1842, p. 7b.

8. H. Le Guern, *Rosoline, ou les mystères de la tombe* (Paris, 1834).

9. H. Le Guern, *Danger des inhumations précipitées* (Paris, 1844).

10. For an example, see H. G. Du Fay, *Des vols d'enfants et des inhumations d'individus vivants* (Paris, 1847).

11. J.-S.-E. Julia de Fontenelle, *Recherches médico-légales sur l'incertitude des signes de la mort* (Paris, 1834).

12. For these stories, see ibid., pp. 101–2 and 106–7, respectively.

13. R. Ferguson in *Quarterly Review* 85 (1849): 346–99.

14. L. Lenormand, *Des inhumations précipitées* (Macon, 1843). The story of Richmodis von Aducht is on p. 91 and three versions of the Two Young Lovers on pp. 34–50, 50–51, and 60–65.

15. Three traditional versions of the Careless Anatomist tale are given in ibid., pp. 32–34; the novel one is on pp. 51–52.

16. For these two stories, see Lenormand, *Des inhumations précipitées*, pp. 27–30 and 52–58, respectively.

17. For this interesting piece of information, see B. Gaubert, *Les chambres mortuaires* (Paris, 1895), pp. 89–90.

18. J.-A. Josat, *De la mort et de ses caractères* (Paris, 1854).

19. This was pointed out by P. Schleisner, *Bidrag til Belysning af Asphyxien og Døden* (Copenhagen, 1868), pp. 14–16, quoting the *Comptes Rendus de l'Academie des Sciences* 35 (1852): 914.

20. W. B. Kesteven in *British and Foreign Medico-Chirurgical Review* 15 (1855): 71–77.

21. For the 1866 and 1869 appeals, see Gaubert, *Les chambres mortuaires*, pp. 93–96.

22. Cardinal Donnet's speech was reported in *Catholic World* 3 (1866): 805–10, and quoted by Tebb and Vollum, *Premature Burial*, pp. 109–13.

23. Charles Dickens's article was in his *All the Year Round*, n.s. 2 (1869): 109–14.

24. These appeals are detailed by Gaubert, *Les chambres mortuaires*, pp. 104–23.

25. See Bouchut, *Traité des signes de la mort*, pp. 5–14, for the Careless Anatomist and pp. 21–28 for the newspaper stories. In the 1883 edition of his book, pp. 28–90 and 183–84, Bouchut also attacked the traditional cases from Bruhier's books, which had been repeated by many later authorities.

26. See the article by E. Bouchut in *Paris Médical* 9 (1884): 133–35.

27. P. Brouardel, *Death and Sudden Death* (London, 1902), p. 40.

28. A. van Hasselt, *Die Lehre vom Tode und Scheintode* (Brunswick, 1862). This work was originally published in Dutch.

29. Ibid., pp. 70–73. Professor Göppert's investigation was quoted from Froriep's *N. Notizen*, no. 23 (1857).

30. Schleisner, *Bidrag til Belysning af Asphyxien og Døden*, p. 4.

31. These three articles are in the *Union Médical*, 3d ser., 30 (1880): 680–81, 785–86, 909–10. They were commented on in a Danish journal, *Ugeskrift for Laeger*, 4th ser., 3 (1881): 36–38, by a doctor who stated that similar loose stories, with as little factual foundation, were often met with in the Danish countryside. Comparable occurrences in England were described in *Lancet*, 1884, vol. 1:968, 1058, and 1908, vol. 1:1431, and in the *Medical Press and Circular*, n.s. 87 (1909): 259.

32. For the Laurens and Lee cases, see Tebb and Vollum, *Premature Burial*, pp. 45, 188.

33. See the articles by H. W. Ducachet in *American Medical Recorder* 5 (1822): 39–53, and N. M. Schreck in *Transylvania Journal of Medicine* 8 (1835): 210–20.

34. See the article by J. G. Kennedy in *Studies in the American Renaissance* (1977) 165–78, for a useful bibliography of early American fictional work on premature burial.

35. See *Littell's Living Age* 13 (1847): 357–58.

36. A. Wilder, *The Perils of Premature Burial: A Public Address Delivered before the Members of the Legislature, at the Capitol, Albany, New York* (London, 1895). This pamphlet was translated into Swedish with the title *Faran för att lefvande begrafvas* (Lund, 1897). Wilder was still active as an anti-premature-burial campaigner in 1900, as judged by an alarmist article of his in *Eclectic Medical Journal* 60 (1900): 388–96.

37. M. R. Fletcher, *Our Home Doctor* (Boston, 1883).

38. E. F. Bishop, *Human Vivisection of Sir Washington Irving Bishop, the First and World-Eminent Mind-Reader* (Philadelphia, 1889). The quotation is on pp. 14–15; quotations from the press about Bishop's death are on pp. 29–54.

39. The strange life and macabre death of Washington Irving Bishop has been admirably described by Ricky Jay in *Learned Pigs and Fireproof Women* (London, 1987), pp. 156–89; see also the article by A. Wilder in *Eclectic Medical Journal* 60 (1900): 388–96, T. J. Hudson, *The Law of Psychic Phenomena* (London, 1893), pp. 309–20, and Tebb and Vollum, *Premature Burial*, pp. 274–75.

40. F. Hartmann, *Buried Alive* (Boston, 1895), reprinted as *Premature Burial* (London, 1896).

41. *British Medical Journal*, 1896, vol. 1:540.

CHAPTER 9: THE FINAL STRUGGLE

1. See the *Burial Reformer* 3 (1913): 75–76.

2. William Tebb's eventful life has been described by M. R. Leverson, *Vaccination in the Light of the Royal British Commission* (Philadelphia, 1900), pp. 57–81. For two samples of his antivaccination books, see W. Tebb, *The Results of Vaccination* (London, 1887) and *Brief Story of Fourteen Years' Struggle for Parental Emancipation from the Vaccination Tyranny* (London, 1894).

3. J. F. Banton, *Vaccination Refuted* (Cleveland, 1882), pp. 9–14. On antivaccinationists in general, see the articles by M. Kaufman in *Bulletin of the History of Medicine* 41 (1967): 463–78, and D. and R. Porter in *Medical History* 32 (1988): 231–52.

4. W. Tebb, *The Recrudescence of Leprosy and Its Causation* (London, 1893).

5. The sinister Dr. Chew's tales are detailed by W. Tebb and E. P. Vollum, *Premature Burial and How It May Be Prevented* (London, 1905), pp. 82–83, 118–20. 124–25, 155, 362–63.

6. W. Tebb and E. P. Vollum, *Premature Burial and How It May Be Prevented* (London, 1896).

7. Tebb and Vollum, *Premature Burial* (1905 ed.), pp. 173–74.

8. Ibid., p. 383.

9. Some of these reviews are reproduced ibid., pp. 29–36; others are in *Perils of Premature Burial* 3 (1912): 28.

10. See J. Stenson Hooker and E. P. Vollum, *Premature Burial and Its Prevention* (London, 1911), for a sample of celebrity comments; others were liberally sprinkled throughout *Perils of Premature Burial* 2 (1909): 32, and 3 (1911): 15, and elsewhere, and in the society's advertisements.

11. These articles were in the *Spectator* 75 (1895): 332, 399, 520, 638–39.

12. J. R. Williamson in *Scientific American* 74 (1896): 294. Williamson was active for many years and wrote letters to various periodicals, warning against the horrors of premature burial. See the *Sanitary Record*, n.s. 31 (1903): 567, and *Perils of Premature Burial* 3 (1912): 31.

13. J. F. Baldwin in *Scientific American* 75 (1896): 315. For other reactions from the medical establishment, see *British Medical Journal*, 1895, vol. 1:730,787–88 and 1896, vol. 1:540.

14. *Medical News* 67 (1895): 657–59.

15. W. M. Weidman in *Lehigh Valley Medical Magazine* 11 (1900): 151–64.

16. *Journal of the American Medical Association* 52 (1909): 1859.

17. D. Walsh in *Medical Press and Circular*, n.s. 54 (1897): 286–89, 316–18, 340–44. This article was later published in pamphlet form as *Premature Burial: Fact or Fiction* (London, 1897). An American edition appeared a year later.

18. *Medical Press and Circular*, n.s. 54 (1897): 413–14.

19. These anecdotes of Lizzy Lind-af-Hageby's agitation are from *Burial Reformer* 1 (1906): 35, 45–48; 1 (1907): 64, 80.

20. Lizzy Lind-af-Hageby became president of the Animal Defence and Anti-Vivisection Society as well as editor of its journal, the *Anti-Vivisection Review*. Much of the early success of the British antivivisection movement was due to her ceaseless agitation and considerable talents for organization. Her ready wit and quick repartee were apparent in her debates with some senior proponents of vivisection, from which the medical grandees often retreated in confusion. In both world wars, she ran hospitals for sick and wounded animals. She continued her work in favor of animal welfare until her death in 1967. For an account of her career see *Who Was Who* 6 (1979): 678.

21. On poor Miss Oakes's disheartening experiences, see *Perils of Premature Burial* 3 (1911): 1–3, 3 (1912): 49–50, and 3 (1913): 78.

22. For these three articles, see *Burial Reformer* 1 (1905): 3–5, 11, 24.

23. This speech was detailed in *Burial Reformer* 1 (1906): 45–48.

24. Hadwen was several times in trouble for his antivaccinationist ideas. He was debarred from membership in the British Medical Association because of unprofessional advertising. He was once charged with manslaughter after a child had died under mysterious circumstances while under his care, but found not guilty. See B. E. Kidd and M. E. Richards, *Hadwen of Gloucester* (London, 1953). Another medical member of the London Association for the Prevention of Premature Burial was Dr. L. A. Parry, who read a paper entitled "The Possibility of Premature Interment" before the Brighton and Sussex Medico-Chirurgical Society in 1914. See *Medical Magazine* 23 (1914): 11–23.

25. On the private bill to prevent premature burial, see *Burial Reformer* 1 (1906): 37–39, and 2 (1908): 6, 9. A similar attempt had been made in the State of New York in 1899, but without success; see the article by H. G. Chapin in *Medico-Legal Journal* (1899): 1–9.

26. R. P. Ferreres and J. B. Géniesse, *La mort réelle et la mort apparente* (Paris, 1905). A new edition appeared three years later. An American edition of this book, entitled *Death Real and Apparent in Relation to the Sacraments* (St. Louis, 1906), lacked Géniesse's valuable additions.

27. These five stories were in *Burial Reformer* 1 (1908): 100, 2 (1908): 4, 3 (1911): 6, 1 (1906): 53, and 1 (1907): 71, respectively.

28. See the *Perils of Premature Burial* 3 (1912): 28.

29. See the London *Times* of July 24, 1926, p. 14e, July 26, 1926, p. 14d, March 31, 1927, p. 16f, Oct. 31, 1928, p. 10d, and Feb. 5, 1936, p. 9a.

30. E. Conner in *Herald of Health* 19 (1895): 189–90.

31. *Medical Press and Circular* n.s. 87 (1909): 259.

CHAPTER 10: LITERARY PREMATURE BURIALS

1. These two stories are in G. Boccaccio: *Il Decamerone: One Hundred Ingenious Novels* (London, 1712), 2:233–39 and 1:153–59, respectively.

2. Bandello's short story, and the later variations on this theme, are discussed by H. Hauvette, *La morte vivante* (Paris, 1933).

3. T. Amory, *The Life of John Buncle, Esq.*, vols. 1–2 (London, 1756).

4. M. Lévy, *Le roman "Gothique" anglais* (Paris, 1995), pp. 413, 632.

5. Mrs. Showes, *The Restless Matron*, vols. 1–3 (London, 1799).

6. J. P. Richter, *Siebenkäs*, in *Werke*, vol. 2 (Munich, 1959).

7. L. A. von Arnim and C. Brentano, eds., *Des Knaben Wunderhorn*, vols. 1–3 (Heidelberg, 1806–8).

8. A. L. Oehlenschläger, *Eventyr, Fortaellinger, Noveller og Roman*, ed. F. L. Liebenberg (Copenhagen, 1861), pp. 93–107.

9. On Poe's notions of death and (premature) burial, see J. G. Kennedy, *Poe, Death and the Life of Writing* (New Haven, 1987), and the articles by R. Mayer in *Poe Studies* 29 (1996): 1–8, and J. E. Pike in *Studies in American Fiction* 26 (1998): 171.

10. It has been suggested by B. K. Brown in *ANQ* 8 (1995): 11 that John Snart's *Thesaurus of Horror* (London, 1817) was one of Poe's sources for "The Premature Burial," but the evidence for this is not impressive. One of his alleged sources, the "Chirurgical Journal of Leipsic," is invented; there was no periodical by that (or any similar) name at the period in question.

11. The *Philadelphia Casket* of September 1827, according to A. H. Quinn, *Edgar Allan Poe* (Baltimore, 1998), p. 418. Ludicrously, C. Quigley, *The Corpse: A History* (Jefferson, 1996), p. 246, reports this case as a true story.

12. "The Buried Alive," in *Blackwood's Magazine* 10 (1821): 262–64; this article was later re-used by several other periodicals in the 1820s and 1830s. See the articles by L. King in *University of Texas Studies in English* 10 (1930): 128–34, and J. G. Kennedy in *Studies in the American Renaissance* (1977): 165–78.

13. See the article by Kennedy in *Studies in the American Renaissance* (1977): 165–78.

14. See the article by W. T. Bandy in *American Literature* 19 (1947): 167–68.

15. This story was quoted by R. Ferguson in *Quarterly Review* 85 (1849): 364, and many other nineteenth-century writers, among them William Hone in his *Every-Day Book*, vol. 2 (London, 1830), pp. 981–82. It was reproduced as "The Dead Alive" in the *New-York Mirror and Ladies Literary Gazette* 4 (1826): 26–27, according to the article by Kennedy in *Studies in the American Renaissance* (1977): 165–78.

16. A. A. Brown in *Nineteenth-Century Literature* 50 (1996): 448–63.

17. F. Kempner, *Denkschrift über die Nothwendigkeit einer gesetzlichen Einführung der Leichenhäusern* (Breslau, 1856; new enlarged ed., 1867). The Swedish translation was entitled *Om Skendöd och Likhus* (Stockholm, 1857); the quotation is from p. 34 of this edition.

18. A good edition of Kempner's poems is *Friederike Kempner, der schlesische Schwan*, ed. G. Mostar (Heidenheim, 1953). The translation of "The Prematurely Buried Child" is slightly abbreviated but shares the spirit of the original. On her biography, see the articles by G. H. Mostar in *Frankfurter Hefte* 7 (1952): 692–700, M. Krohn in *Jahrbuch der Schlesichen Friedrich-Wilhelms Universität zu Breslau* 8 (1962): 233–46, and T. V. Laane in *Dictionary of Literary Biography* 129 (1993): 174–81.

19. Gottfried Keller's poem "Lebendig begraben" was written in 1846, but not published until 1883, in his *Werke*. See I. Stoessel, *Scheintod und Todesangst* (Cologne, 1983), pp. 88–89. On Keller's life and works, see L. Gessler, *Lebendig begraben* (Basel, 1964), and J. M. Lindsay, *Gottfried Keller* (London, 1968).

20. See Stoessel, *Scheintod und Todesangst*, pp. 117–21, and E. Vogl, *Der Scheintod, eine medizingeschichtliche Studie* (Munich, 1986), pp. 20–28, which give sources to relevant biographical works.

21. On Gogol, see G. Prause, *Genies ganz privat* (Düsseldorf, 1975), p. 324, and Vogl, *Der Scheintod*, pp. 26–27. C. Boetzkes, in *Scheintot begraben* (Percha, 1984), pp. 139–40, declares himself convinced that Gogol was buried alive, but the coffin might just have been tilted when put in the grave, or the body might have turned over during the process of putrefaction.

22. J. S. Le Fanu, *Ghost Stories and Mysteries*, ed. E. F. Bleiler (New York, 1975), pp. 1–93.

23. Marie Corelli, *Vendetta* (London, 1886). On the novel's publication history, see B. Masters, *Now Barabbas Was a Rotter: The Extraordinary Life of Marie Corelli* (London, 1978), pp. 61–69, 123.

24. W. Collins, *Jezebel's Daughter* (London, 1880); see also C. Peters, *The King of Inventors: A Life of Wilkie Collins* (London, 1992), p. 397.

25. S. L. Clemens, *The Complete Short Stories of Mark Twain* (New York, 1957), pp. 225–34. The story "A Dying Man's Confession" was originally in his *Life on the Mississippi*.

26. A. Dumas, *The Count of Monte Cristo* (London, 1898). Another novel by Dumas *père, Pauline, or Buried Alive* (London, n.d.), features a woman deliberately entombed in a family vault by her evil husband.

27. G. de Maupassant, *Tales of Supernatural Terror*, ed. A. Kellett (London, 1972), pp. 43–48, 74–79. See also Stoessel, *Scheintod und Todesangst*, pp. 90–92.

28. E. Zola, "La mort d'Olivier Bécaille" (1884), translated into English in *Little Masterpieces of Fiction*, ed. H. W. Mabie and L. Strachey, vol. 7 (New York, 1904), pp. 17–57.

29. S. Lagerlöf, *From a Swedish Homestead* (London, 1901).

30. On Andersen, see Stoessel, *Scheintod und Todesangst*, pp. 118–19.

31. On Nielsen, see the article by O. Hagelin in *Bokvännen* 434 (1996): 19–25.

32. On Nobel, see Stoessel, *Scheintod und Todesangst*, pp. 118–19.

33. F. Rolfe's "How I was Buried Alive" appeared in *Wide World Magazine* 2 (1898): 139–46.

34. H. Lewis, *The Haunted Husband* (New York, 1899).

35. G. Atherton, *The Bell in the Fog, and Other Stories* (London, 1905).

36. A. Conan Doyle, *His Last Bow* (London, 1917), pp. 199–233.

37. H. P. Lovecraft, *Selected Letters*, vol. 4, (Sauk City, Wisc., 1976), letter 371.
38. J. Dickson Carr, *The Three Coffins* (New York, 1935).
39. C. Woolrich, *Nightwebs*, ed. F. M. Nevins Jr. (London, 1973), pp. 3–52.
40. D. Wheatley, *The Ka of Gifford Hillary* (London, 1956).
41. M. Crichton, *The Great Train Robbery* (London, 1976).

CHAPTER 11: WERE PEOPLE REALLY BURIED ALIVE?

1. J. Snart, *Thesaurus of Horror* (London, 1817), pp. 27–28.
2. [Anon.], *Uppmaning till Menniskovännen . . .* (Uddevalla, 1840).
3. H. Le Guern, *Danger des inhumations précipitées* (Paris, 1844).
4. L. Lenormand, *Des inhumations précipitées* (Macon, 1843).
5. Dr. de Pietra Santa in *Journal d'Hygiène* (Paris) 22 (1897): 157–63.
6. B. Gaubert, *Les chambres mortuaires* (Paris, 1895).
7. J. G. Ouseley, *Earth to Earth Burial* (London, 1895).
8. J. Stenson Hooker and E. P. Vollum, *Premature Burial and Its Prevention* (London, 1911).
9. Quoted by G. E. Mackay in *Popular Science Monthly* 16 (1880): 389–97.
10. M. Dana in *Arena* 17 (1897): 935–39.
11. E. P. Vollum in *Undertaker's Journal*, April 1904, quoted by R. Davies, *The Lazarus Syndrome* (London, 1999), p. 132.
12. T. M. Montgomery in *The Casket* (Rochester, N.Y.), March 2, 1896, quoted in W. Tebb and E. P. Vollum, *Premature Burial and How It May Be Prevented* (London, 1905), pp. 81–82.
13. Hebenstreit's observations are reviewed in A. van Hasselt, *Die Lehre vom Tode und Scheintode* (Brunswick, 1862), pp. 65–66.
14. Dr. von Röser's observations were published in Sachs's *Medizinische Jahrbücher* of 1859, quoted ibid.
15. T. K. Marshall in *Medicolegal Journal* 35 (1966): 14–24.
16. F. J. Holzer in *Deutsche Zeitschrift für die gesamte gerichtliche Medizin* 62 (1968): 95–100.
17. P. Hedenius in *Upsala Läkareförenings Förhandlingar* 10 (1874–75): 1–45, pp. 35–36.
18. These three cases were quoted in Tebb and Vollum, *Premature Burial* (1905 ed.), pp. 84–87.
19. On the *Totenlaut*, see E. Vogl, *Der Scheintod, eine medizingeschichtliche Studie* (Munich, 1986), p. 53.
20. The Bobin case was described by Tebb and Vollum, *Premature Burial* (1905 ed.), pp. 55–56, from an article in the *Hereford Times* of November 16, 1901. It had also reached the *Gazette Médicale de Paris*, ser. 11, vol. 2 (1899): 581, where it was accepted as one of live burial.
21. Van Hasselt, *Die Lehre vom Tode und Scheintode*, pp. 70–73. The case was in the *Nederlandse Weekbl. v. Geneeskundigen* 1854, p. 252.
22. This case was described by Dr. Löscher in *Vierteljahrschrift für gerichtliche und öffentliche Medizin* 14 (1858): 170–72.
23. For information about the *Sarggeburt* phenomenon, see also the later reviews by Jungmichel and Musick in *Deutsche Zeitschrift für die gesamte gerichtliche Medizin* 34

(1940–41): 236–56, and M. Wagner, *Die Bedeutung des Scheintodes aus rechtsmedizinischer Sicht* (Munich, 1982), pp. 14–15.

24. V. Starcke, *Giertrud Birgitte Bodenhoffs Mysterium* (Copenhagen, 1954); V. Møller-Christensen in *Medicinsk Forum* 6 (1953): 184–91.

25. E. Bodenhoff, *Den Gamle General* (Copenhagen, 1914), pp. 10–13, and N. Lange, *Den Bodenhoffske Slægtebog* (Copenhagen, 1914), p. 63.

26. A. Fabritius in *Personhistorisk Tidsskrift*, ser. 13, vol. 3, no. 3 (1955): 1–19.

27. See the articles by H. Schirmer in *Tidsskrift for den Norske Lœgeforeningen* 94 (1974): 2135–39, K. Lithner in *DAST-Magazine* 21, no. 4 (1988): 4–10, and M. H:son Holmdahl in *Saga och Sed* 11 (1990): 35–43.

28. C. Pfendler, *Quelques observations pour servir à l'histoire de la léthargie* (Paris, 1833).

29. M. Rosenthal in *Medizinische Jahrbücher herausgegeben von der K. k. Gesellschaft der Aerzte* 2 (1872): 389–99.

30. T. M. Madden in *Medical Magazine* 4 (1897): 857–62, 922–29.

31. C. H. Miles in *Medical Times and Hospital Gazette* 33 (1905): 423–24.

32. S. Weir Mitchell, *Diseases of the Nervous System*, 2nd ed. (London, 1895), pp. 189–90.

33. G. Gilles de la Tourette, *L'hypnotisme et les états analogues au point de vue médico-légal* (Paris, 1887). See also the discussion by P. Vinge in *Norsk Magasin for Laegevidenskaben*, ser. 4, vol. 4 (1889): 486–502, 558–75, 829–41.

34. On cataplexy, see the articles by C. Guilleminault et al. in *Archives of Neurology* 31 (1974): 255–61, J. G. van Dijk et al. in *British Journal of Psychiatry* 159 (1991): 719–21, and C. Guilleminault and M. Gelb in *Advances in Neurology* 67 (1995): 65-77.

35. Dr. von Jäger's paper was in the *Zeitschrift für die Staatsarzneikunde* 6 (1823): 241–52.

36. A. Devergie in *Annales d'Hygiène* 34 (1870): 310–27.

37. See P. Brouardel, *Death and Sudden Death* (London, 1902), pp. 26–27, 40, and the article by G. Puppe in *Deutsche medizinische Wochenschrift* 46 (1920): 383–85.

38. See the articles by Puppe in *Deutsche medizinische Wochenschrift* 46 (1920): 383–85, and J. Fog in *Ugeskrift for Lœger* 92 (1930): 11–15.

39. Wellcome Institute collection of images, no. 30192–3.

CHAPTER 12: ARE PEOPLE STILL BEING BURIED ALIVE?

1. The tale of Essie Dunbar is recounted by P. Sieveking in *Fortean Times* 49 (1987): 55–60, quoting the Augusta, Ga., *Gazette* of Aug. 25, 1955, and Dec. 5, 1985.

2. On Angelo Hays, see an article in the German magazine *Stern* (1979, no. 13), pp. 80–82; also C. E. Boetzkes, *Scheintot begraben* (Percha, 1984), pp. 9–16, and E. Vogl, *Der Scheintod: Eine medizingeschichtliche Studie* (Munich, 1985), pp. 50–51.

3. For these three versions, see, respectively, Boetzkes, *Scheintot begraben*, pp. 9–16, S. Schäfer, *Scheintod: Auf den Spuren alter Ängste* (Berlin, 1994), pp. 64–68, and J.-Y. Péron-Autret, *Buried Alive* (London, 1983), pp. 63–64.

4. Boetzkes, *Scheintot begraben*, pp. 145–47.

5. Schäfer, *Scheintod*, p. 8.

6. P. Sieveking, "Misidentification of the Dead and Premature Burial, 1991–1997" (MS), p. 4.

7. See the articles by E. Rautenberg in *Deutsche medizinische Wochenschrift* 45 (1919): 277–78, and G. Puppe, ibid., 46 (1920): 383–85. C. Quigley, *The Corpse: A History* (Jefferson, 1996), p. 186, is wrong to claim that Minna Braun's coffin was actually buried.

8. The article by M. Duvoir and L. Pollet in *Bulletin de la Société Médicale des Hôpitaux de Paris*, 3d ser., 50 (1934): 801–5, is an early example of ECG recordings in cases of apparent death.

9. See the articles by D. J. Powner in *Journal of the American Medical Association* 236 (1976): 1123, and P. M. Black in *New England Journal of Medicine* 299 (1978): 338–44.

10. *Läkartidningen* 68 (1971): 11–15.

11. H. A. Edwards et al. in *British Journal of Anaesthesiology* 42 (1970): 906–8.

12. *La Presse Medicale* 78 (1970): 1291, 1585.

13. Sieveking in *Fortean Times* 49 (1987): 55–60.

14. This case was reported in the Swedish newspaper *Dagens Nyheter* on March 16, 1988.

15. *New York Post*, June 16, 1993, and other newspaper articles.

16. *Daily Telegraph*, March 5, 1996, and other newspaper articles.

17. On this technique, see the articles by P. Husby et al. in *Intensive Care Medicine* 16 (1990): 69–72, and B. H. Walpoth et al. in *European Journal of Cardiothoracic Surgery* 4 (1990): 390–93.

18. See T. Koch, *Lebendig begraben* (Augsburg, 1996), p. 193, and the article by H. de Varigny in *Revue Generale des Sciences Pures et Appliqués* 43 (1932): 367–71.

19. G. Rossow in *Beiträge zur gerichtlichen Medizin* 17 (1943): 121–26.

20. Péron-Autret, *Buried Alive* (London, 1983). The original French edition was published in 1979.

21. Boetzkes: *Scheintot begraben.*

22. No fewer than eighteen examples of alarmist German magazine and newspaper articles are given by Vogl, *Der Scheintod*, pp. 71–85.

23. See the articles by H.-J. Mallach et al. in *Medizinische Welt* 28 (1977): 1905–8, H.-J. Mallach in *Deutsches Ärzteblatt* 78 (1981): 893–98, U. Stirner in *Medizinische Welt* 32 (1981): 1460, W. Eisenmenger et al. in *Beiträge zur gerichtlichen Medizin* 40 (1982): 49–53, and H.-F. Brettel and H. von Lüpke, ibid., 42 (1984): 359–62. The thesis by E. Vogl, *Der Scheintod*, specifically describes the recrudescence in Germany of the fear of *Scheintod* and takes issue with the alarmist arguments of Boetzkes and others.

24. *Newsweek*, Dec. 18, 1967, p. 87.

25. Boetzkes, *Scheintot begraben*, pp. 115–32; also Vogl, *Der Scheintod*, pp. 86–88. The coroner's inquest was reported in the *Times* (London) of Feb. 27, 1974, p. 1g, and March 16, 1974, p. 1h.

26. See the articles by J. B. Posner in *Annals of the New York Academy of Sciences* 315 (1978): 215–27, P. M. Black in *New England Journal of Medicine* 299 (1978): 338–44, C. Pallis and B. MacGillivray in *Lancet*, 1980, vol. 2:1085–86, and C. Pallis in *British Medical Journal* 286 (1983): 123–24.

27. B. Jennett et al. in *British Medical Journal* 282 (1981): 533–39, and other studies reviewed by D. Lamb, *Death, Brain Death, and Ethics* (London, 1988), pp. 63–69.

28. See the article by C. F. Bolton et al. in *Lancet*, 1976, vol. 1:535.

29. Lamb, *Death, Brain Death, and Ethics*, pp. 63–64.

30. See the editorial comment by the distinguished neurologist Dr. C. Pallis in *British Medical Journal* 281 (1980): 1084. A number of letters from outraged medical practitioners who had watched the *Panorama* documentary were published ibid., 1139–41.

31. See the article by B. A. Bradley and P. M. Brooman in *Lancet*, 1980, vol. 2:1258–59.

32. M. Sanner in *Journal of the American Medical Association* 271 (1994): 284–88.

33. Schäfer, *Scheintod*, is knowledgeable on the older German literature, but credulous with regard to the old and new mythology on the subject, as well as lacking in medical expertise. This author also maintains a Web site (*www.scheintot.de*). R. Davies, *The Lazarus Syndrome* (London, 1999), lacks both medical and historical insights. For recent examples of irresponsible alarmist magazine and newspaper articles, see those by T. Grove in *Daily Mail*, Sept. 4, 1999, pp. 28–29, and P. Michalski, *Bild-Online*, Sept. 9, 1999.

34. C. Pallis and D. H. Harley, *ABC of Brainstem Death* (London, 1996), provides a valuable overview of these matters.

35. In the section on these newspaper myths, I rely heavily on the archive of newspaper clippings collected by Mr. Paul Sieveking, editor of the *Fortean Times*, and on a manuscript of his entitled "Misidentification of the Dead and Premature Burial, 1991–1997."

36. For these three modern versions of the Careless Anatomist, see *Independent on Sunday*, Jan. 14, 1996, *People* Sept. 11, 1994, and *Celebrity*, March 31, 1988, respectively.

37. See the article by P. Sieveking in *Fortean Times* 63 (1992): 36–38; the story originated in an AFP news bulletin.

38. See the articles by P. Sieveking in *Fortean Times* 49 (1987): 55–60, and 63 (1992): 36–38.

39. See P. Sieveking in *Fortean Times* 63 (1992): 36–38, quoting a 1901 instance from Kansas; for a later, dramatic account, see *Weekly World News*, Sept. 24, 1991.

40. For some of these silly reports of Italian and Australian resurrections in various British newspapers, see *Sun*, Aug. 23, 1992, *Sunday Sport*, Sept. 10, 1992, *Star*, Jan. 11, 1993, *Today*, April 19, 1994, and *Daily Star*, Jan. 5, 1998.

41. *Weekly World News*, Aug. 30, 1988.

42. Sieveking, "Misidentification of the Dead and Premature Burial, 1991–1997," p. 2.

43. *News of the World*, April 26, 1992, and *Daily Record*, Dec. 24, 1994.

44. *Sunday Mail* (Scotland), Dec. 31, 1995.

45. Sieveking, "Misidentification of the Dead and Premature Burial, 1991–1997," p. 1.

46. For the two tales of "Mum's fatal move," see *Daily Mirror*, March 30, 1992, and *Daily Star*, Sept. 15, 1995. Paul Sieveking has exposed a similar recycling operation, concerning two elderly Italians leaping out of their coffins, in the *Fortean Times* 63 (1992): 36–38.

47. Sieveking, "Misidentification of the Dead and Premature Burial, 1991–1997," p. 2.

48. *International Daily Express*, May 11, 1999.

49. J. H. Brunvand, *Curses! Broiled Again!* (New York, 1996), pp. 66–68.

50. See *New York Post*, Nov. 18, 1994, and many other U.S. newspapers.

51. See *Daily Mail*, July 18, 1996, p. 35, and many other UK newspapers. The case was investigated by Davies, *The Lazarus Syndrome*, pp. 55–57, which found that Maureen Jones was still alive and reproduced her photograph.

52. E. Morselli, *Sulla dismorfofobia e sulla tafefobia* (Geneva, 1891).

53. L. Löwenfeld, *Die psychischen Zwangserscheinungen* (Wiesbaden, 1904), p. 121.

54. S. Feldman in *Psychiatric Quarterly* 16 (1942): 641–45.

55. H. Dietrich in *Schweizer Archiv für Neurologie, Neurochirurgie und Psychiatrie* 120 (1977): 195–203.

56. Personal communication from Ms. Holly Stadtler, Dream Catcher Films.

57. *Notes and Queries*, 3rd ser., 2 (1862): 110.

58. See, in particular, I. Stoessel, *Scheintod und Todesangst* (Cologne, 1983), pp. 102–121, and Vogl, *Der Scheintod,* (Munich, 1986), pp. 19–22.

59. Other safeguards of course include embalming and cremation. According to J. J. Farrell, *Inventing the American Way of Death, 1830–1920* (Philadelphia, 1980), p. 163, fear of premature burial was one of the three causes of the rise in embalming in the United States, along with preservation and disinfection.

INDEX

Note: Page numbers in *italics* refer to illustrations.